WICKED WEATHER

WICKED WEATHER

Natural Disasters that Changed History

ALEXANDER GATES

BLOOMSBURY ACADEMIC
NEW YORK • LONDON • OXFORD • NEW DELHI • SYDNEY

BLOOMSBURY ACADEMIC
Bloomsbury Publishing Inc, 1359 Broadway, New York, NY 10018, USA
Bloomsbury Publishing Plc, 50 Bedford Square, London, WC1B 3DP, UK
Bloomsbury Publishing Ireland, 29 Earlsfort Terrace, Dublin 2, D02 AY28, Ireland

First published in the United States of America 2026

Cover design by Diana Nuhn
Cover image: © EyeEm Mobile GmbH

A catalog record for this book is available from the Library of Congress.

ISBN: HB: 978-1-5381-9893-3
ePDF: 979-8-8818-5829-2
eBook: 978-1-5381-9894-0

Typeset by Deanta Global Publishing Services, Chennai, India
Printed and bound in the United States of America

CONTENTS

1

What Are Natural Disasters?

What are natural disasters? Some might think that natural disasters are things like all tornadoes and all earthquakes. However, there are earthquakes every day of the year, and most go unnoticed. Similarly, the United States experiences about 1,200 tornadoes per year, and most of them do no damage.[1] Both earthquakes and tornadoes can be natural disasters, but they must cause death and destruction on a significant level to be considered natural disasters. In fact, natural disasters are catastrophic loss of life and/or property caused by natural events.[2] This includes biological threats such as disease, injurious bacterial colonization, invasive dangerous plants, and insect infestations, among others. This book, however, focuses solely on natural disasters involving the physical threats from the earth, hydrosphere, and atmosphere. These disasters include earthquakes, volcanoes, tsunamis, hurricanes and typhoons, tornadoes, droughts, and floods. One difference between physical threats and biological threats is that humans can do nothing to control the physical threats, whereas they may have or develop some control over the biological threats. All people can do to avoid physical threats is to shelter in safe locations, engineer structures to withstand the impact, and evacuate areas before the threat produces a disaster.

Phenomenal efforts and huge amounts of money are invested in medications and medical procedures to increase human life spans by even a short amount of time.[3] These efforts have increased the average life span in many countries and protect people from many natural threats. This increased protection has led to the expectation and entitlement of security among many to the point of not worrying about the threats in the developed world. These increased expectations of security have given many people the courage to inhabit areas that may be too prone to natural threats to be safe. The sense of entitlement is not the only reason that people live in unsafe areas. Some people are stuck in unsafe areas because they lack the finances, resources, or ability to relocate to safer areas. In either case, many people in many parts of the world are at risk for injury or death from natural disasters.

In terms of earth history, the geological and meteorological events that produce natural disasters are not extraordinary.[4] Like the earthquakes and tornadoes, they occur daily and have done so since the beginning of time. Many impressively large events go unnoticed because they don't impact humans. Unlike the way primitive cultures regarded these events as being of supernatural origin as a punishment for wrongdoings,[5,6] they are part of the natural sequence of events. That is one part of natural disasters. The other part is the impact on humans and the response of society to these natural events. Few people want to experience a natural disaster. However, many people are mesmerized by them and marvel at their destructive power. In an attempt to find reasons for disasters, some cultures attempt to explain them through folklore and religious beliefs. Scientists seek to understand the processes behind them in an attempt to ultimately forecast them and reduce their destructiveness.

There have been several natural disasters that changed the course of history in profound ways by themselves.[7] The natural event coupled with the impact on people and society makes these natural disasters complex issues. In any evaluation of natural disasters, both aspects are considered. The controlling science is the first part of analyzing a natural disaster. The societal aspect cannot be generalized because each situation is different. The best way to appreciate the full meaning of a natural disaster is through case studies of the most impactful events.

Considering the profound destructiveness of natural disasters, the question is why anyone would want to live in a natural disaster-prone area? The geographic areas dominated by specific types of natural disasters are well-known. Some areas are even known by their common natural disasters like "Tornado Alley" in the midwestern United States or the "Ring of Fire" rimming the Pacific Ocean. There are worldwide systems to warn people of impending disasters, so residents can make preparations to escape harm. In addition to evacuation, these preparations include advanced engineering solutions that reduce the impact of the disaster. Having such disaster preparations sharply contrasts with most of human history, where people were completely unprepared for natural disasters and surprised by them. Only modern technology has changed this situation by reducing its impact. However, humans have made no progress in preventing natural disasters, and there is no indication that they ever will.

There are two primary reasons why people now inhabit natural disaster-prone areas in such abundance. The first is that the human population has ballooned from about 1 billion in 1800 to 7.7 billion today.[8] As a result, urban areas are sprawling into less safe locations to accommodate this growth. Communities now encroach on the damage zones of recent volcanic eruptions, inviting natural disasters. For example, the 79 AD (CE) eruption of Mount Vesuvius flattened the Roman cities of Pompeii and Herculaneum, killing an estimated 20,000 people. Despite this threat, the population of the city of Naples is nearly 3.1 million, and the inhabited area now encroaches on the eruption area of the volcano to a far greater degree than it did in ancient Roman times.[9]

The populations of many cities have increased so much that destructive events that formerly were not catastrophic now are. The city of New Orleans,

Louisiana, has been struck by major hurricanes throughout its history, yet Hurricane Katrina in 2005 was more destructive than all previous storms.[10] The population and urban sprawl had increased dramatically since the previous major hurricane, which resulted in many more casualties. A powerful earthquake struck Port-au-Prince, Haiti, in 2010 that killed between 100,000 and 316,000 people, making it one of the deadliest natural disasters in history.[11] The reason that it was so devastating is that the population of Port-au-Prince boomed to more than 10 million people with no precautions for earthquakes despite it having suffered earthquakes in 1770 and 1842 that destroyed the city twice and a tsunami that devastated the harbor in 1946.[12]

There are also psychological reasons why more people are living in natural disaster-prone areas. Humans increasingly attempt to control nature and have succeeded in very significant ways. If a natural disaster has not struck a prone area for as short as several decades, residents become complacent and assume they are safe. They quickly overdevelop areas as the market demands that are marginally safe and occupy them with an attitude of entitlement and impunity. The city of Miami, Florida has grown dramatically in recent decades with large buildings right out on the beaches. This development is in spite of a major hurricane having destroyed the city about a century ago in 1926.[13] It is estimated that if a similar storm strikes the city again, it will cause damages of more than $150 billion in 2005 ($235 billion today) and be the costliest natural disaster in US history. In 1928, the even more powerful Okeechobee hurricane struck Florida just north of the Miami area and killed more than 2,800 making it the second deadliest hurricane in US history. In 1992, Hurricane Andrew caused extensive damage to an area just west of Miami. It was the costliest hurricane in the United States at the time causing more than $27 billion in damage, and it held the record for more than twenty-five years. Despite the potential for damage and loss of life in Miami from a major hurricane, fear does not appear to be slowing its popularity.

The final major reason that people live in natural disaster-prone areas is that they cannot leave. Dangerous areas that are prone to flooding, landslides, insect swarms, or other threats tend to be less expensive and therefore more commonly occupied by impoverished people who cannot afford to live in more expensive and safer areas. Certain racial and ethnic groups may also be forced into areas that are less safe for economic factors or discrimination. An example of this occurred during the 2005 Hurricane Katrina disaster in New Orleans, Louisiana. Evacuation orders were issued for residents of New Orleans well ahead of the storm making landfall. Virtually all of the residents who had the capacity to flee the city did so in a mass exodus. When the storm finally struck, an overabundance of the residents left to face its fury were from underserved minority groups, living in poverty or near poverty and/or too old to flee. They inordinately suffered the most death, injuries, and hardships when the disaster ensued.[14]

Relocating to less dangerous areas may also be hampered by family, cultural, and employment constraints. People could have family obligations that keep them in dangerous areas even if they have the finances to move away. These could involve caring for older parents and relatives or the need to be near family for childcare. People may also have careers that keep them in dangerous areas such as port workers or emergency workers. Comfort in a community or the "we've always lived here" attitude could also keep residents in an unsafe area even if they can afford to move.

Death Tolls

Death tolls are typically regarded as the most important measure of the intensity of a natural disaster. With low numbers of dead, the residents consider themselves lucky in contrast to those experiencing higher numbers. However, the numbers may not always tell the whole story.[15] If the death toll numbers are low and the disaster is geographically constrained, the death toll can be considered relatively accurate in most cases. Death tolls in large events are commonly less accurate because multiple sources of casualties are compiled to reach the total, which introduces multiple sources of error. In ancient events, survival was the main concern making the counting of the bodies much less of a priority. As a result, the death toll in ancient events could be in error by as much as a tenfold or more. Death tolls for recent events tend to be more accurate in most cases because the handling of disasters is better organized, and data collection is more meticulous. However, even in recent events, some disasters are prone to more counting errors. In inland earthquakes where rescue and relief efforts are organized and well planned, the numbers of casualties tend to be more accurate. In the cases of earthquakes, the collapsed structures remain in place as do the bodies, so as excavation proceeds, counting is relatively accurate. In coastal floods like hurricanes, typhoons, and tsunamis, there can be large errors in the numbers of casualties even today. With the sweeping in and withdrawal of such large amounts of seawater, bodies can be swept out to sea or at least displaced along with houses and even whole neighborhoods. Municipal buildings including halls of records may even be washed away, removing all residency and birth records of residents. In these cases, numbers of casualties tend to be very inaccurate. Casualty records can be equally inaccurate in major volcanic eruptions as a consequence of the destruction caused by the intense explosions and engulfing lava flows.

Improper translations of accounts of natural disasters from other languages may also lead to errors in casualties and damage, which can then be disseminated throughout the literature. For example, one of the worst-reported earthquake disasters in history occurred in China in the 1920s. Most normally reliable information sources report a death toll for this event of 200,000 people. A recently released, more reliable translation reports that 200,000 livestock perished in this event, but only 35,000 people died. It was still a horrible disaster, but much less of one than is widely reported.

Societal Response to Natural Disasters

In ancient civilizations, bad timing or location of natural disasters could result in the downfall of cities and empires. If two armies were battling or even threatening each other and a natural disaster weakened one of the groups, it could decide the outcome of the conflict. Severe earthquakes, tsunamis, storms, and volcanic eruptions have been responsible for several of the great historical catastrophes that changed history. This is well exemplified in the eastern Mediterranean Sea region where activity tends to be episodic. Large earthquakes in the region tend to occur regularly over short 50–100-year intervals, which are punctuated by long periods of quiescence that can last many hundreds of years. The active periods of multiple earthquakes are termed "seismic crises."[16]

Most ancient cities were protected by large encircling walls made of stacked rock in arrangements with variable stabilities. The shaking from a large earthquake could cause sections of them to collapse, breaching them and making the city vulnerable to attack. Such attacks could be from (i) external enemies during wars such as Joshua at Jericho and the Arab attack on Jerusalem in 31 BCE, (ii) neighboring adversaries in long-term stand-off conflicts like Mycenae's fall in 1200 BCE and Saul's battle around 1020 BCE, and (iii) the uprising of enslaved peoples like the Helots in Sparta in 464 BCE, the Hattusas around 1200 BCE, and Teotihuacan around 700 CE. If a small earthquake struck during a minor or modest conflict, damage might take a few tens of years to fully repair with a minor shift of power, if any. However, if a severe earthquake struck during a major military conflict, it might take hundreds of years to rebuild the damage, and the city would be plunged into a dark age during that time with all efforts on survival rather than rebuilding. During these incidents, the outcome of a conflict could be decided by the earthquake, and therefore, radical shifts in power were possible. Cities have even been abandoned if the damage was too severe or the people were too overwhelmed by the conflict or resulting hardship.

Underreporting of Casualties

For these reasons, residents of impacted regions downplayed the extent of damage and casualties from natural disasters in ancient societies in many cases. This practice of downplaying the impact of disasters existed throughout most of human history and is not completely gone today. The deadly 1976 Tangshan earthquake in China still has an official death toll of 242,000 people. However, recent studies by unbiased international researchers suggest that the death toll was probably closer to 655,000 people, making it potentially the worst natural disaster in history.[17] However, this earthquake occurred during the height of the Cold War, and the communist government at the time did not want the country to look weakened by the disaster to its enemies. As a result, a deflated death toll may have been purposefully reported.

A good example of downplaying or even ignoring a natural disaster for political reasons occurred with the 1949 Khait Earthquake in the USSR. The powerful earthquake struck the Gharm Oblast region of now Tajikistan on July 10, 1949, at a strong magnitude of 7.4–7.5. The earthquake was disastrous on its own, but the majority of the damage and loss of life was caused by the massive landslides it generated.

The area of the earthquake in eastern Tajikistan includes the Pamir Mountains, which are known as the "Roof of the World" because of their height and the high Pamir Plateau. The steep slopes generate regular rockfalls, landslides, flash floods, and avalanches even without the shaking from an earthquake. However, the area contains complex plate tectonic interactions that produce strong earthquakes adding to the threat. The Khait earthquake occurred at the southern edge of the Tien Shan belt, which extends westward from China. The Khait earthquake is interpreted to have resulted from movement on the major Vakhsh fault. In 1911, an earthquake in this area generated a large landslide that blocked an entire river valley, which consequently flooded it and created the huge earthquake Lake Sarez, which was 36 miles (60 km) long and at least 1,625 feet (500 m) deep. This event attracted the interest of American seismologists, and they sought to conduct research in the area. Through an unprecedented cooperative study with Soviet seismologists at the time, the international team established a seismograph network around the region and a research lab in the town of Gharm on the Surkhob River.[18]

Heavy rains in July 1949 soaked the soils of the steep slopes in the area. Then, earthquake activity increased, which caused minor landslides on many of the soggy slopes. When the Khait earthquake struck, the stage was already set for massive landslides. The intense shaking caused hundreds of small landslides to initiate, and they coalesced into a major landslide with an estimated volume of 320.4 million cubic yards (245 million m^3). This huge mass traveled 12.4 miles (20 km) in the Yasman Valley down a very shallow slope of 2°. In the nearby Yarhich and Obidara-Khauz valleys, the landslides began as rockslides that entrained water-saturated soils. The main slide traveled 4.6 miles (7.41 km) while descending 4,662 feet (1,421 m) in elevation. This steep slope accelerated the slide to an average speed of 97.5 feet (30 m) per second, but in some areas, it achieved a harrowing speed of 325 feet (100 m) per second. Estimates of the volume of this slide are as high as 327 million cubic yards (250 million m^3). These landslides and avalanches made a deafening noise, generated strong winds and raised huge dust clouds that encased the entire region.

The death and destruction from this natural disaster were staggering. Every building in an area 37–41 miles (60–65 km) long and 4–5 miles (6–8 km) wide was destroyed by the earthquake and landslides. The entire town of Khait and at least twenty neighboring towns were buried in soil and debris to a depth of as much as 100 feet (30 m). In all, about 150 towns and villages were impacted. The burial of so many towns made determining the death toll of the disaster difficult to impossible. The minimum number of deaths is about 7,200 people, but a more realistic death toll is between 12,000 and 28,000 people.[19]

The Khait earthquake occurred at the beginning of the Cold War during the rule of Josef Stalin. The USSR regarded disasters as a sign of weakness that could encourage their enemies to attack them. Consequently, the government effectively covered up and disavowed the Khait disaster, not admitting to any fatalities for many years. The true extent of the disaster was not revealed until the past decade.[20]

Today, between the towns of Gharm and Jirgital, Tajikistan, a large monument of white marble commemorates the tragedy. The statue depicts a woman with her head down in grief, and it overlooks the Yarhich River Valley where the largest landslide burial remains unexcavated. It is a grim reminder of the power of earthquakes and landslides and the danger inherent in people living in the Pamir.

Possible Inflation of Casualties

Modern world politics have changed the response of governments to natural disasters in many nations. In recent times, many of the wealthier countries began providing aid to countries struck by natural disasters. Even private groups are now providing significant aid. In 1970, ex-Beatle George Harrison developed a new idea of a benefit concert to help Bangladesh, which had been devastated by the Bhola typhoon.[21] It generated attention for the people suffering in the country as well as an enormous amount of aid to help relieve it. The success of this effort encouraged artists and promoters to organize many enormous benefit concerts both for natural disasters and for suffering of any kind on a worldwide basis. A new level of benefit concerts was achieved in 1985 when the LiveAid concert to address the terrible famine in Ethiopia was broadcast worldwide to an estimated 1.9 billion people. It included most of the top names in Rock music and generated enormous amounts of aid. This new attitude compelled countries to more accurately report the number of casualties in a natural disaster or even to inflate them, possibly to receive more aid. A possible case of casualty inflation occurred with the 2010 major Port-au-Prince earthquake, Haiti.

On January 12, 2010, a major earthquake struck Haiti with an epicenter 15 miles (25 km) west-southwest of the city of the Port-au-Prince[22] as discussed earlier. The magnitude of the event was a powerful 7.0. The island of Hispaniola is seismically active, experiencing earthquakes that destroyed Port-au-Prince in 1751 and 1842 and a recent earthquake of magnitude 6.8 in the Dominican Republic in 1985. Despite this background of earthquake disasters, the city was ill-prepared for the powerful event. The earthquake was produced by movement on a fault related to the Enriquillo-Plantain Garden fault system, which also produced large earthquakes in 1860, 1770, and 1751. This fault system is one of two that cross the island.[23] These faults are part of the boundary that separates the Caribbean plate from the North American plate. This tectonic plate boundary connects the Caribbean arc and the Central American arc. The movement on this plate boundary and faults is about 0.8 inches (2.0 cm) per year. It has produced numerous other earthquakes including the 1692 Port Royal and 1907 Kingston earthquakes in Jamaica and the 1972 Motagua earthquake in Guatemala that killed at least 23,000 people.[24]

Seismic waves from the earthquake produced extreme ground shaking in Port-au-Prince and its suburbs. Shaking was felt throughout the region including in Cuba, Jamaica, Venezuela, Puerto Rico, and especially in the bordering Dominican Republic. This shaking devastated the city of Port-au-Prince. The reason it was so devastating is that it was the worst earthquake in over 200 years in Haiti, which, as a consequence, was not prepared for earthquakes. The country is the poorest in the Western Hemisphere, where 80 percent of the people live in poverty and more than 50 percent live in abject poverty. Haiti had substandard building codes before the earthquake, and few of the builders followed them anyway. Most buildings barely survived normal conditions, much less shaking from a strong earthquake. The collapsing buildings caused extreme loss of life. The United Nations estimates that 105,000 homes were destroyed and 208,000 were damaged, mostly in Port-au-Prince. The earthquake also produced a tsunami in the Port-au-Prince harbor that killed at least four people.[25]

It is estimated that more than 3,000,000 people were directly impacted by the earthquake. The official Haitian government estimates are that 316,000 people were killed, 300,000 were injured, and 1,000,000 were made homeless. If the 316,000 deaths are correct, it would be the worst modern earthquake disaster and the second worst ever. However, the death toll was later calculated by unbiased international researchers to be between 92,000 and 160,000 people[26] or half or less of the official total. The government refused to revise its total. The Inter-American Development Bank estimated that the total cost of the disaster was between $7.8 billion and $8.5 billion, based on a death toll of between 200,000 and 250,000 people.

The international response to the earthquake disaster began immediately. On January 13, 2010, the United Nations Secretary-General Ban Ki-Moon released a statement about the humanitarian disaster in Haiti. United States president Obama pledged $1.15 billion toward rescue and relief. Venezuela pledged $1.32 billion in aid, and the International Monetary Fund canceled Haiti's $268 million debt. Many other countries also made large pledges and sent troops, rescue personnel, and supplies. Former president Clinton and the Haitian prime minister led a twenty-six-person team to oversee the distribution of the $5.3 billion raised in the first two years after the earthquake. Many countries and benefit fund raising efforts continued to generate relief funding for many years. Using an aggressive rebuilding plan, the Haitian government solicited nearly $10 billion in pledges at a March 2010 donor conference. The high death toll estimates made the request for aid more compelling. It is questionable why the Haitian government retains the high death toll claim despite evidence to the contrary.

Complete Misrepresentations

Sometimes through the propagation of reporting errors by mistake or purposefully, natural disasters have been reported that never actually happened. The ancient natural disasters have a higher likelihood of being fabrications because they are difficult to verify, but even more recently reported events have the possibility of never having happened. The 1737 Calcutta, India earthquake appears near the top of a number of supposedly reputable lists of the world's greatest natural disasters.[27] In these lists, it is reported to have a death toll of up to 300,000 people, which ranks it among the top three or four natural disasters of all time. More recent analyses of this disaster, however, not only cast doubt on the astounding death toll but also on whether there was even an earthquake at all. The confounding issue is that on October 11, 1737, the day of the reported earthquake, a major cyclone also struck the Calcutta area. This storm was so intense that it sunk eight of the nine British ships stationed in the area. It also sunk virtually all of the boats owned by the natives that were moored in the Ganges River at the time. There is no doubt that this storm caused many deaths.

The complicating factor was that the British East India Company, which operated in the area, did not submit a report of the disaster until survivors returned to the headquarters in London more than six months later. At that point, their report listed about 3,000 fatalities from the storm, but there was no mention of an earthquake at all. Despite this report, the following year, stories in both The Gentleman's Magazine and The London Magazine reported that some 300,000 people died in a devastating earthquake with no mention of a cyclone. This contradiction caused the confusion about the event and eventually the earthquake became the favored interpretation.

A major problem with the earthquake interpretation is that historical records indicate that, in 1737, the urban population of Calcutta was less than

20,000 people. Admittedly, there were other settlements in the Calcutta area, but records indicate that there were likely no more than 300,000 people in the entire region. However, there could not have been that many people killed. Further, although the Calcutta area has been sporadically struck by small to moderate earthquakes, there is no record of a single large event like those that strike the Himalayan region to the north on a regular basis. For these reasons, it is less likely that such a powerful earthquake would strike Calcutta. It is most likely that the 1737 Calcutta earthquake was not an earthquake at all[28] much less a record-breaking natural disaster.

What Can People Do to Stay Safe?

There are many ways for people to ensure their safety from natural disasters. The most obvious of these is not to live in places that are struck by frequent natural disasters. This would mean that many inhabited areas would need to be abandoned, and, even then, it would not guarantee full safety from catastrophic events. The reason for this is that disasters can occur even in unexpected areas. The best solution in most cases is to avoid the frequently struck areas but otherwise plan for the events that cause disasters by developing engineering solutions and early warning and evacuation systems so people can stay safe through any natural event that might strike.

The simplest solution for many disasters is an early warning system.[29] These systems typically consist of public education about the potential events, loud sirens, television and radio announcements, and cell phone alerts. In frequently struck areas, there may be marked escape routes for evacuation and emergency drills in middle and high schools. The challenging aspect of warning systems is the detection of an impending disaster far enough in advance for residents to act. Volcanoes near populated areas are well monitored using advanced technology, and warnings are typically issued well ahead of the eruptions.

Although they are severely destructive, in most cases, people evacuate the impacted areas well ahead of the eruption. Hurricanes, cyclones, and typhoons typically move slowly enough that they can be accurately tracked by national weather services, and their paths are predicted. In many cases, residents can be safely evacuated or are able to shelter in place ahead of the danger. Far fewer people are impacted by these events than would be without the systems. The problem with these storms is that they can exhibit unexpected tracks and strengths, and they can strike areas that cannot be evacuated in time, such as islands, so many people are still injured and killed when they strike. Tsunamis can also be tracked, and, in many cases, the systems provide ample warning for evacuation. However, this is not always the case especially in areas where they are uncommon. It typically takes a monumental disaster before an effective early warning, education and evacuation system is fully implemented. On the other hand, tornadoes strike so quickly and with such fury that they are commonly not as easy to identify or predict by national or local weather services. In many cases, there is little warning and people must take shelter in a home tornado shelter or in a reinforced, safer part of a house, if at all. At this point, earthquake events cannot be predicted to within decades or centuries.

The other way people can protect themselves from natural disasters is through engineering solutions. Each solution is different depending on the threat. The construction practices to withstand natural threats continue to improve in industrialized countries as the result of updated building codes and new technologies. For example, California updates its building codes after each large earthquake. This is a proactive policy against this type of natural disaster. Even the pyramid-like, iconic TransAmerica building in San Francisco was constructed with that shape to be earthquake resistant. Most large and modern buildings in that city do not transmit seismic waves into the structure through engineering advancements that are now required building practices. In addition to excellent earthquake engineering, Japan has the most advanced tsunami engineering. There has been installation of walls and gates in many critical coastal locations to protect them from tsunami waves. When a tsunami is sensed by the sophisticated monitoring system in the Pacific Ocean, steel gates are slid shut in low-lying areas to prevent most waves from penetrating inland and wreaking havoc. The Netherlands developed among the most advanced technologies of walls, gates, pumps, and canals to protect them from flooding.[30] They are highly prone to flooding with much of the country at or even below sea level.

These are examples of advanced disaster reduction engineering, but there are many smaller and simpler methods as well.[31] In the mountains of Peru, the ancient complex of Machu Picchu includes earthquake reduction techniques in its construction, in which rocks slide against each other to absorb seismic waves. In Sicily, towns built rock walls along the slopes of the Mount Etna volcano to obstruct the regular lava flows from reaching inhabited areas. In disaster-prone areas, many public buildings are constructed with extra reinforcement to serve as emergency shelters during natural disasters. Residents are alerted to evacuate to these buildings during emergencies. In the midwestern United States, many residents construct tornado rooms, shelters, or cellars that they can retreat to during a tornado. Local building codes may even include requirements for disaster reduction in home construction.

How effective can building codes be at reducing deaths and destruction during a natural disaster? A great example is the comparison between similar earthquakes in Haiti and California. California updates their building codes after every strong earthquake, so construction continues to improve. In 1989, the Loma Prieta (World Series) earthquake struck California with a strong magnitude of 6.9 just 44 miles (70 km) from the city of San Francisco.[32] It produced very strong shaking throughout the city and was very similar in character to the 2010 Port-au-Prince, Haiti earthquake. Even the faults that generated the two earthquakes are similar, both being dominated by lateral rather than vertical movement and both separating tectonic plates. However, this earlier powerful earthquake killed only sixty-three people, and forty-one of them died in a single highway collapse.

The impacts of the 1989 Loma Prieta earthquake are in stark contrast to those of the 2010 Port-au-Prince earthquake. About 5,000 times more people perished in the Port-au-Prince earthquake than in the Loma Prieta event. This comparison strikingly demonstrates the control that construction practices play in preventing destruction and loss of life during natural disasters. However, California, and especially San Francisco, is one of the most affluent areas in the world, and the community can afford to engineer their buildings to best survive earthquakes. Haiti, on the other hand, is one of the poorest countries in the world and cannot afford to engineer its buildings to withstand natural disasters. Therefore, the impact, death toll, and destruction in natural disasters depend on location as much as the power of the event at least in recent history.

This means that two events can have the same physical conditions, the same type of geological or meteorological setting, similar population densities and distribution in the impacted area, and timing during the day and year, and yet one could change history and the other could just be a serious event. This, however, only applies to modern events of the past fifty to sixty years when disaster reduction engineering began in earnest. In fact, geology students in the 1960s were often shown a film on earthquakes entitled "San Francisco: City that waits to die," which details how unprepared San Francisco was for a large earthquake. It took a concerted effort and many years to change that prognosis. This is not to say that San Francisco cannot still be destroyed by a large enough earthquake, but clearly, it is much better prepared than it was.

This also means that two events in the same location and all other conditions being essentially the same but from two spaced time periods will likely not have the same outcome. San Francisco was also struck by a large earthquake in 1906.[33] The Great San Francisco earthquake was at least ten times as powerful as Loma Prieta with an estimated magnitude of 7.7–7.9 so not quite comparable, but it caused the deaths of about 3,000 people, making it the second worst natural disaster in US history. This is a huge difference in impact, and it did change history to some degree. However, perhaps the deadliest part of the disaster was the fires it caused, which razed the city and caused much of the death and destruction. In 1906, most heating, cooking, and lighting were accomplished with open fires, which were knocked over by the shaking from the earthquake. Fires can still occur in earthquakes, but they are generally minor and contained. So the other generality about natural disasters that can be made is that timing is almost as important as location.

2

What Makes Natural Disasters Occur in Certain Areas?

Certainly, the systems that control the natural events that can cause natural disasters are complex, which makes predicting their timing or even occurrence difficult. Further, there are enough anomalous occurrences of these events that it is impossible to say that any single area can't be struck by some event that could cause a natural disaster. For example, the populated northeastern corridor of the United States is not prone to any events that could cause a natural disaster yet the New York metropolitan area was struck by a severe drought and a sizable earthquake in 2024,[1] and it was engulfed by hazardous levels of wildfire smoke in 2023.[2] It was struck by a hurricane that produced a natural disaster in 2011 and again in 1938 which also devastated Connecticut, Rhode Island, and Massachusetts.[3] A "tornado alley" type F3 tornado struck western Massachusetts in 1953.

There are even stranger anomalies on a global scale. In 2023, the arid Libyan desert of North Africa was struck by a massive flood that killed at least ~6,000 people and as many as 14,000.[4] This natural disaster was generated by an anomalous hurricane-like storm that formed in the Mediterranean Sea. Scorching Saudi Arabia was struck by a snowstorm for the first time ever in 2024.[5] So certainly, anomalies make completely ruling out natural disaster-producing events besides volcanic eruptions nearly impossible in most populated areas. However, there are general earth systems that generate the great preponderance of disaster-producing events in certain areas. These areas are commonly struck by natural disasters, and no one is surprised when they are. This chapter will describe why these areas are so prone with the understanding that anomalies are possible.

The Atmosphere

There are three types of natural disasters described that originate from the atmosphere: temporary climate shifts or aberrations, hurricanes, and tornadoes. The aberrations in climate can cause either abnormally dry periods or abnormally wet periods. The designation of "abnormal" depends on the location and the normal climate conditions. Dry spells are expected in the Sahara Desert, but if they occur in a tropical rainforest or even a temperate area like most of Europe or the eastern United States, they can be catastrophic.

The atmosphere is the layer of gases held to the earth by gravity. The density of gases is highest closer to the planet surface, and they are progressively less dense with higher elevation. There are subdivisions in the atmosphere based on gas density and other conditions, and they include the troposphere, stratosphere, mesosphere, thermosphere, exosphere, and outer space. The troposphere is from the surface to about 7.5 miles (12 km) in elevation. It contains all of the air used for respiration and photosynthesis and 99 percent of the water vapor and aerosols. Virtually all surface weather takes place in the troposphere with only minor amounts extending into the stratosphere.[6]

One control on the weather is that warmer air holds more moisture. Humidity is the measure of this moisture contained in the air. When moist warm air is cooled, the moisture condenses into water droplets. A glass of a cold drink appears to sweat on a humid summer day as moisture in the air condenses on the cold surface. A second control is that the troposphere decreases in temperature with elevation. This causes mountains to be snowcapped while the lower parts are not. The air also holds less moisture at higher elevations than in the lowlands because of the colder temperatures. Air pressure also decreases with altitude. The air pressure around us at about 14 pounds per square inch or 1 bar of pressure at sea level. People are used to the air pressure where they live. However, if they travel from low altitudes to the mountains, they notice how "thin" the air is and may experience labored breathing. The vertical movement of air also impacts air pressure. Sinking air increases the pressure at the surface (high pressure), whereas rising air decreases the air pressure (low pressure). Upward movement of air experienced in low pressure carries the warmer, moister air of the surface to the colder high altitudes. The cold causes the moisture in the rising air to condense, and precipitation occurs. Downward movement of air creating high pressure carries colder, drier air to lower altitudes where it is warmer and can hold more moisture, resulting in clear skies.

Climate Disaster: Droughts and Floods

These processes control local weather patterns and the large-scale weather patterns of the earth. The large-scale, long-term weather patterns constitute the climate. The main driver of the climate system is the air rising at the equator because it is heated by the sun and becomes less dense. It then spreads out both north and south as it rises to the upper troposphere. In contrast, the cooling of incoming upper troposphere air at the poles makes it denser, and it sinks pulling elevated cold air toward the surface. This rising air at the equator spreads poleward at high altitude, and the sinking air at the poles moves toward the equator along the surface as prevailing winds and forms a large-scale convection cell.

This simple pole-to-equator air convection system is complicated by the Coriolis Effect. This effect is caused by the earth rotation. The relative speed on the surface from rotation depends on the position on the planet. At the equator, a person travels all the way around the earth or 24,000 miles (38,624 km) in one day of twenty-four hours relative to a fixed point. That means they would be traveling at 1,000 miles per hour (1,609 kph). In contrast, a person standing at the north or south pole would just spin one full turn in the same twenty-four-hour period which is basically a speed of 0 miles per hour. In general, all points closer to the equator always move faster than those poleward.

The Coriolis Effect twists the simple pole–equator circulation system and breaks it into three convection cells per hemisphere called the Hadley, Ferrell, and Polar cells, from the equator to the poles. The rising air at the equator generates low pressure, condensation, and precipitation. The tropical rainforests occur in the equatorial region worldwide for that reason. The Hadley cells drive cold, dry air downward toward the surface at 30° latitude. This produces high pressure with clear, dry, and calm weather there. For this reason, most deserts of the world are concentrated here including the Sahara, Arabian, and Australian deserts. The next poleward cells produce rising air at 60° latitude, producing low pressure and consequently the temperate rainforests. The air falls at the poles, which produces a desert. The poles don't appear to be deserts because they are covered in by ice and snow. However, these ice deposits took many thousands of years to accumulate.

In addition to the climates defined by the ascending and descending bands of air, local climates and microclimates develop in response to surface features. Oceans modulate the local temperatures and humidity, and they are especially impactful if they contain strong warm or cold currents. Oceans can produce local disturbances that change expected to abnormal precipitation patterns. An abundance of landmasses can reduce evaporation to the atmosphere, which changes what the climate model dictates. This situation will change what should be rainy areas into much drier areas. Siberia is within the northern-temperate rainforest zone, but it is dry because there is so much landmass instead of the ocean to provide moisture. Local topography can also change the local climate. In the northwestern United States, moist air from the Pacific Ocean is driven eastward and up the west side of the Cascade Mountain range. Moving air up a mountain is the same as lifting it in low pressure. As a result, the west side of the Cascade Range is a temperate rainforest from this orographic precipitation, but now, the descending dry air on the east side of the range produces a desert in eastern Washington.

Any deviation from the normal patterns created by these climate systems can cause a devastating natural disaster. Even a few months of little or no rain in a normally wet area can cause a significant drought. If it lasts longer than that, there may not be enough water for animals, vegetation, and humans, which, in turn, can cause a famine. If these conditions persist, a natural disaster will ensue. Similarly, if an area becomes abnormally wet, floods can result causing a natural disaster. Shifting ocean currents can cause these extreme weather patterns. El Niño and La Niña weather phenomena cause radical changes in temperature, precipitation, and severe weather patterns including natural disasters on a global scale. The cause of these phenomena is shifts in Pacific Ocean circulation patterns. El Niño results in severe floods in Peru and devastating droughts and wildfires in Indonesia and Australia. It also causes major changes in global atmospheric circulation. The most devastating proven El Niño in history lasted from 1876 to 1878 and caused massive droughts that triggered disastrous famines in Asia, Africa, and South America and killed more than 50 million people globally or about 3 percent of the world population.[7]

Tornadoes

When a mass of air sits above an area for enough time, it takes on the characteristics of that area. If the air mass develops over central Canada, it is cold and dry. If an air mass sits over the Gulf of Mexico, it is warm and moist. Over the North Pacific, it is cold and moist. These air masses can move and interact with other air masses. The boundary between any two air masses is called a front. There are several different types of fronts depending upon the nature of the colliding air masses. There are cold fronts, warm fronts, and occluded fronts, but only a cold front produces the extreme weather that can cause a natural disaster.

Cold air replacing warm air in an area indicates the passage of a cold front. In the United States, cold air masses typically move from northwest to southeast or, less commonly, inland from the Pacific or Atlantic Oceans. Cold air masses from the north and northwest are drier than the air masses they contact. Cold air is denser than warm air, so it drives under warm air at the cold front. The front is very steeply sloped, which makes the weather at the cold front intense but short-lived. Cold fronts are more common in the summer and display thick clouds that build up quickly with heavy rain and strong winds. Thunderstorms are common, and tornadoes are possible along these fronts.[8]

Thunderstorms require vigorous vertical air circulation, which is common in strong cold fronts. The upward motion carries warm moist air to great heights where it condenses producing heavy rain and hail, and the downward motion produces surface winds and possible funnel clouds and tornadoes. The strengths of the upward versus downward motions determine the type and degree of danger produced by the storm. There can be a single circulation cell in the thunderstorm, which is called a single-cell storm. Two or more circulation cells are called multicellular storms, which are more powerful than single-cell storms and can produce tornadoes. Large storms with especially vigorous circulation are called supercells, which produce virtually all major tornadoes and tornado outbreaks and are responsible for the vast majority of natural disasters.

The United States has the ideal weather to produce tornadoes, and it is struck by about 1,200 tornadoes per year as a result.[9] Cold dry air from

Canada sweeps southeastward to the area east of the Rocky Mountains, where it collides with the warm, moist air that streams northward from the Gulf of Mexico. Cold fronts form where these two air masses collide, and they can be very active. The collision zone cold front can lie anywhere from Florida west to Texas, north to Minnesota, east through Ohio, and south to the Carolinas. Typically, the severe weather migrates from the Gulf Coast in the spring to the Midwest later in the summer. If the two air masses are different enough and other favorable conditions prevail, an outbreak of multiple tornadoes can occur, which is extremely dangerous. If just the right conditions of profoundly different air masses and a slow-moving front occur, a superoutbreak may be produced where hundreds of tornadoes can be generated over a day or two.

Hurricanes, Typhoons, and Cyclones

Hurricanes and their equivalents of cyclones and typhoons are among the deadliest natural disasters in the world. The worst natural disaster in US history was the Galveston Hurricane of 1900, which killed more than 8,000 people. The worst natural disaster of the twentieth century was the 1970 Bhola cyclone of Bangladesh which claimed between 300,000 and 500,000 victims. Even with modern, sophisticated weather prediction technology, hurricanes can still devastate even prepared areas. Hurricanes are cyclones in the Indian Ocean and typhoons in the Pacific Ocean. All are produced through the same processes.[10]

Hurricanes do not form at fronts but instead within large, stable air masses. They begin as weakly developed clusters of thunderstorms in the Atlantic Ocean off the coast of West Africa, possibly from the uplift of air through heating over the Sahara desert. To thrive and grow larger, they require a surface ocean temperature of 79 °F (26 °C) or greater. At those temperatures, the humidity becomes elevated in the air above the ocean surface. This warm, moist air is swept by the updrafts in the thunderstorms. As this warm, moist air rises, the moisture in it condenses, forming rain. The condensation from vapor in the air to liquid water releases latent heat energy into the storm. This energy powers up the storm, causing more vigorous circulation and stronger updrafts. The enhanced updrafts continually lower the air pressure in the storm. This decreased pressure sweeps more air into the storm, producing ever-stronger surface winds. The strong winds whip up the surface of the ocean into ever larger waves that expose greater ocean surface area from which more humidity can be emitted. More humidity drives more condensation and greater latent heat transfer, all of which quickly power up the storm. This powering-up process continues as the disturbance transitions from a tropical depression to a tropical storm to a hurricane as it drifts across the Atlantic Ocean.[11]

Because they are not associated with fronts, hurricanes have no steering mechanisms and can slowly drift in unpredictable paths. They can sit still for long periods of time and rain excessive amounts, causing extreme flooding. They can also accelerate along incoming fronts to as much as 60 miles per hour (100 kph), driving extreme winds and storm surge along the coast. If high-level wind shear develops, the storm will weaken. They can also undergo rapid intensification, strengthening from a survivable Category 1 to a Category 5 monster in as little as twenty-four hours.[12] These erratic tendencies make hurricane paths and impacts difficult to predict.

The need for temperatures of 79 °F (26 °C) or greater to intensify the developing storm requires that hurricanes, typhoons, and cyclones can only form in the tropics but over any ocean worldwide. They also can only form at warmer times of the year, which in the northern hemisphere is from about June to October. A well-developed storm can cause catastrophic destruction and death. If the storm gets strong enough and the steering currents cooperate, these storms can drift outside of the tropics and devastate areas that are not used to them. Once they make landfall, they quickly degrade but they can still drop deadly levels of flood waters in areas that can't handle them.

Hurricanes can strike throughout the Caribbean Sea, in Central America, around the Gulf of Mexico, in the southeastern United States northward through the Carolinas, and possibly in northernmost South America. They can be powerful and cause catastrophic damage and loss of life. The worst recent storm was Hurricane Mitch in 1998, which had a death toll of 22,000 people in Honduras and Nicaragua.[13] Typhoons drift across the broad expanse of the Pacific Ocean. They can strike Hawaii and other islands, but many focus their energy on the Philippines as their first landfall. These storms can cause massive damage like Supertyphoon Haiyan, which may have killed more than 10,000 people in 2013.[14] Typhoons commonly make a second landfall in Taiwan, Japan, China, or Southeast Asia. These second landfalls can produce excessive rainfall, which can cause disasters on their own. For example, in 1975, Typhoon Nina stalled for several days within China dropping more than 64 inches (163 cm) of rain in a short period of time, making it a once-in-2,000 years flood. This massive amount of water overfilled the reservoir behind the huge Banqiao dam,[15] which collapsed from the water pressure, and the resulting deluge killed about 26,000 directly and as many as 240,000 people indirectly from the disease and famine it caused.

Cyclones form in the Indian Ocean and can be especially deadly if they turn north up the Bay of Bengal. The danger is that much of the coastal region from India through Bangladesh and Myanmar is at sea level. Even a moderate storm surge from a cyclone can be deadly, but large storm surges are catastrophic. Perhaps the worst natural disaster of the twentieth century was the 1970 Bhola cyclone that struck the Bangladesh area, causing the deaths of 300,000–500,000 people in the lowland delta region of the country.[16] However, this was not the only catastrophic cyclone in the area. Another deadly cyclone struck the same area of Bangladesh in 1991 and claimed almost 139,000 victims. The latest catastrophic storm was Cyclone Nargis, which struck Myanmar in 2008, claiming more than 138,000 lives from the Irrawaddy delta region.[17] The death tolls of these Indian Ocean cyclones dwarf the hurricanes and typhoons because of the condition of the areas they strike, not because of the size, speed, or power of the storms.

Geographic Control on Earthquakes, Volcanoes, and Tsunamis

The natural disasters from processes in the earth's lithosphere are primarily controlled by plate tectonics. Rather than the earth being solid underfoot, it must be regarded as flexible and mobile. Through this understanding, the types and locations of volcanoes, earthquakes, and tsunamis can be placed into context. However, the locations of mountains, oceans, continents, island chains, and even the shape of the seafloor can also be understood because of it. Plate tectonics has been called the "glue that holds geology together" because it explains the locations and relationships of most of the geology of the earth.

The earth contains chemical layers and mechanical layers that are distinct but overlapping. The chemically distinct layers are, from outside, the crust, mantle, and core.[18] The core is divided into two mechanical layers: a solid inner core and a liquid outer core. Chemically, the core is composed of an iron–nickel mix, and it produces the earth's magnetic field. The next layer out from the core is the mantle, which is a thick layer of dense iron–magnesium silicate rocks. It is divided into an upper and a lower mantle layer based on the mineral density. The outermost layer of the earth is the crust. There is both continental and oceanic crust. Ocean crust is younger, thinner, and denser than continental crust. Gravity pulls the ocean crust deeper into the planet because of its density. Between its density and its thinness, the elevation is much lower than continental crust. Because gravity also draws water to the lowest point, ocean water covers the ocean crust. Continental crust is thicker and lighter than ocean crust, so it floats up and comprises the landmasses.[19]

The outer mechanical layer of the earth is the lithosphere, which is comprised of ocean and/or continental crust and a thin layer of rigid mantle attached underneath it. The lithosphere occurs in rigid fragments or plates of various sizes, shapes, and compositions. Beneath the plates, the mantle is gum-like, which defines the next lower mechanical layer called the asthenosphere. The depth to the asthenosphere is defined by the thickness of the lithosphere and the topography of its surface. Higher mountains mean greater depth to the asthenosphere. These rigid lithospheric plates float on the asthenosphere. A hard-boiled egg with the shell cracked makes a good visual illustration of this situation. The hard-shell fragments would be like the lithospheric plates. They can be pressed down into the rubbery egg white, which has a similar relative consistency to the asthenosphere, though it is gummier. Lithospheric plates can even be depressed into the asthenosphere by weight on the surface.

However, the analogy with the hard-boiled egg does not explain the movement of the lithospheric plates. The broken shell fragments on the hard-boiled egg don't move, but the seven major and several smaller lithospheric plates on the earth are constantly moving. They move because the asthenosphere is constantly flowing underneath them, and the plates are carried along like big rafts. The flow is driven by hot zones that fuel big thermal convection cells in the asthenosphere, where hotter gummy rock rises upward because it is less dense. As this hot rock reaches the top of the asthenosphere, it spreads away from the hot area in both directions and cools. Cooling causes the gummy rock to get denser, and it sinks down into the deeper asthenosphere by gravity. There, it is reheated and goes through the convection cell process again like a circular recycling system.[20]

Because there are several flowing convection cells around the earth, the lithospheric plates move in different directions and grind against each other. The boundaries where the plates meet are the sources of the vast majority of earthquakes, volcanoes, and tsunamis on earth. There are three types of interactions between the plates at their margins: pushing together or convergence, pulling apart or divergence, and sliding past each other or transform. Convergence results from asthenosphere flows in opposite directions that drive plates into each other in a collision. Divergence results from flows that are away from each other and drives plates away from each other. Transform results from oppositely directed flows that are next to each other, which causes plates to slide past each other laterally. These zones where the plates are interacting are margins or boundaries, and each one produces its own group of events that can result in natural disasters.[21]

Divergent Margins

Two lithospheric plates move in opposite directions away from each other at divergent margins. Beneath them, hotter asthenosphere rises upward along a map-wise linear to curved zone and splits, flowing directly away from it in both directions. Each of these opposite flows is at the top of adjacent convection cells with opposite senses of rotation. Natural disasters are common above the divergent boundaries but only if they are on land. In divergent margins under the ocean, few natural disasters occur, if any.

Divergent boundaries mostly develop on continental crust as a continental rift zone and slowly transition to a mid-ocean ridge in the middle of a major ocean basin. The rift zone likely begins as a line of hot plumes in the asthenosphere under a continent. The plumes coalesce into a linear zone of rising hotter asthenosphere. The flow in the asthenosphere from this line pulls overlying continent apart, producing a continental rift. This causes the bottom of the plate to stretch and thin, but the land above it breaks into a series of faults, which produce significant earthquakes. The increased heat generates magma in the asthenosphere, which rises through the crust above, producing low-lying volcanoes whose eruptions are typically not dangerous. However, the heat from this magma melts the continental rocks, producing a second type of magma. This second magma generates steep-sided, explosive volcanoes that can be very destructive and cause natural disasters.

As the splitting of the continent continues, a depression forms between the two sides and, eventually, the continent splits into two. The depression then floods and becomes a long, narrow sea, which widens to an open ocean, and the divergent boundary is the mid-ocean ridge in the middle of it. Volcanic activity becomes more intense during separation but once it is in the open ocean, it no longer poses a threat. The only exception to this is if the mid-ocean ridge becomes land that can be inhabited, like Iceland, in the Atlantic Ocean. The coasts of open oceans are called passive margins because there is no volcanic activity and little earthquake activity. The Atlantic Ocean is an example of a full ocean basin with passive margins along both sides and a volcanic mid-ocean ridge in the middle.

Convergent Margins

Convergent margins are where lithospheric plates are pushed together in a collision. Two facing convection cells in the asthenosphere form currents that flow toward each other in convergent margins. As the flowing gummy rock spreads away from the hot divergent boundaries, it cools and becomes denser, finally sinking deeply into the asthenosphere at the zones of convergence. Consequently, the plates above them are driven into each other in a collision along linear or arcuate convergent boundaries or margins. There are three types of convergent boundaries depending on the types of crust that are colliding with each other: ocean–ocean, ocean–continent, and continent–continent.

As two lithospheric plates are pushed together, at the convergent boundary, one plate slides under and the other plate rides above. The denser plate slides under, and the lighter plate rides above. Ocean crust is much denser, at about 3.4 gm/cm^3, than continental crust, which has a density of about 2.65 gm/cm^3.

Therefore, where the two plates push together, ocean crust is always driven beneath continental crust forming a subduction zone. Bathymetrically, subduction zones form trenches on the ocean floor, and these trenches are the deepest places on the planet.

As the oceanic plate is driven into a subduction zone, it carries rocks and sediments to great depth, where they are pressurized. Water escapes from these wet rocks and causes the asthenosphere above it to melt, producing magma. This magma is buoyant in the dense asthenosphere around it, which causes it to rise rapidly to the surface where it produces extensive volcanism. On land, the mountain range of volcanoes it produces is called a magmatic arc. The classic magmatic arc is the Andes Mountains of South America. The volcanoes in continental magmatic arcs produce some of the most spectacular and deadly volcanic explosions on earth. Many of them produce natural disasters. As a result, the more dangerous of these in terms of regularity in strong activity and nearby population centers are meticulously monitored, and the nearby public is well educated on evacuation procedures in case of activity. Further, the compressing of the two plates and the bending and breaking of the ocean crust as it is driven into the subduction zone produce strong earthquakes. Many of these are the strongest in history and also produce some of the worst natural disasters on earth. Because these earthquakes occur at the interface with an ocean, they can produce very large tsunamis. The most powerful earthquakes are mostly produced on the subduction zone itself and are termed megathrust earthquakes. The tsunamis they generate are the most damaging in history including the 2004 Indian Ocean tsunami, the 2011 Tohuku, Japan tsunami, and most of the others described in this book.

In other cases, ocean crust is driven under other ocean crust in an ocean–ocean collision. The geometry of these is similar to an ocean–continent collision with a subduction zone and a trench. However, because it forms on ocean crust that has such a low elevation, an arcuate chain of volcanic islands, called an island arc or volcanic arc, forms above the subduction zone such as the Aleutian Islands of Alaska and Caribbean Islands. Similar to the ocean–continent collisions, these zones are marked by intense volcanic eruptions and earthquakes that commonly produce tsunamis. There tend to be fewer people living on the volcanic islands than around continental arcs, so the volcanoes and earthquakes tend to be less devastating locally but the tsunamis are just as dangerous as continent–ocean boundaries. Nonetheless, these zones are extremely dangerous and produce numerous natural disasters.

Many lithospheric plates are composed of a piece of continental crust attached to oceanic crust. As subduction of the ocean crust part of a plate under a facing continental plate continues, the remaining ocean crust is used up. At some point, it runs out, and the attached continent collides with the facing continent on the overriding plate. This situation constitutes a continent–continent collision. An example is the Indian Ocean and Timor Sea north of Australia. Australia is the trailing continent in an ongoing ocean–continent collision between the ocean crust part of the Australian plate and Indonesia. Australia will collide with Indonesia in a continent–continent collision when the ocean crust runs out. The Himalayan Mountains, Zagros Mountains in Iran, and many other major mountain systems of the world were built this way.[22]

As the continent–continent collision proceeds, the trailing continent tries to follow the ocean crust into the subduction zone. However, continental crust is too light to descend into the dense asthenosphere. Instead, the continent on the subducting plate drives under the overriding continent, forming a huge fault between the two. Other faults form, and the crust is compressed and crumpled into a highly elevated and broad mountain range and plateau on the overriding continent. The volcanism stops as soon as the continents make contact, so there is no threat of volcanic disasters. However, the numerous active faults produce regular, very strong earthquakes throughout the area and far into the overriding continent. The Himalayan continent–continent collision is caused by the northern margin of India driving under the overriding southern margin of Asia. The entire seismically active area from this collision extends far into Nepal, China, Pakistan, and Afghanistan, among other areas.[23] The numerous devastating earthquake natural disasters in this part of the world are the result of this continental collision.

Transform Margins

The primary plate margins are divergent and convergent. However, the earth is a sphere which makes the plate boundaries curved and the speed of convection cells in the asthenosphere variable. The misalignment and asymmetry of boundaries or differences in relative plate velocity mean that the plate movement cannot be handled by the straight or slightly arcuate boundaries, and compensation features must develop. These compensation features are called transform margins or boundaries, and they take care of all of the misfitting of plates. Transform margins or boundaries are large faults with primarily lateral motions in contrast to divergent margins where faults drop the surface down into depressions and convergent margins where faults push the surface up to form mountains. As compensation features, transform margins must connect the two other margin types.[24] Therefore, the possible types of transform margins are named by the margins they connect: divergent–divergent, divergent–convergent, and convergent–convergent. These zones are marked by strong earthquakes and resulting natural disasters but only a few minor volcanic eruptions, if any. They also may produce primarily local and minor tsunamis because the fault movement is horizontal whereas large tsunamis are produced by vertical movements.

The most common transform margin is the divergent–divergent type. Almost all of these are on ocean crust, where they compensate for curves and variable spreading rates on mid-ocean ridges. Mid-ocean ridges appear as long submarine mountain ranges on bathymetric maps of the ocean floors of the world, and transforms appear as perpendicular lines that offset them all along their lengths. An example can be seen on the mid-Atlantic ridge, which lies north to south along the center of the Atlantic Ocean. Upon close inspection, hundreds of lines crossing perpendicular to the ridge, and all of the lines are divergent–divergent transform margins.

Convergent–convergent transform margins are much rarer. They most commonly form as offsets on island arcs and are primarily on the ocean floor. There is an excellent example of a convergent–convergent transform margin on land crossing the South Island of New Zealand. This transform is the South Alpine Fault which produces regular major earthquakes. Another convergent–convergent transform crosses from the Caribbean arc to the Central American subduction zone along its west coast. This transform produced the 2010 Port-au-Prince earthquake in Haiti, as well as major earthquakes in Jamaica and Guatemala that produced natural disasters.[25]

Convergent–divergent transforms are also rare. They mostly occur on smaller plates especially where young ocean crust is being subducted. There are several good examples of convergent–divergent transforms like the Dead Sea transform in the Middle East, but the best studied example is the San Andreas Fault of California. The divergent margin is the Gulf of California in Mexico, and the convergent margin is the Cascadia Subduction Zone from northern California through Washington and into Vancouver, Canada. The San Andreas produces regular earthquake activity that causes natural disasters. The 1989 Loma Prieta earthquake and the 1906 Great San Francisco earthquake are just two examples of this activity.

3

Volcanic Eruptions

Asteroid collisions with the earth are the most impactful single-episode natural disasters. These space rocks of 33 feet (10 m)–621 miles (1,000 km) in diameter come hurtling out of the sky at 50,000 miles per hour (80,500 kph) and wreak such havoc upon collision that the entire planet jolts. So much dust is raised that it blots out the sun for an extended period of time, the atmosphere cools dramatically, and vegetation is stunted if it grows at all.[1] The term "nuclear winter" was applied to this scenario in the 1980s when nuclear arsenals were being reconsidered. The Chixulub crater in the Yucatan Peninsula of Mexico marks the asteroid collision that wiped out the dinosaurs 66 million years ago.[2] It is this scale of change that large body extraterrestrial impacts can create. Fortunately, sizable impacts are very rare, and no such natural disasters have been recorded in historical times.

The second most devastating natural disasters are volcanic eruptions. Taken together in a volcanic province, they can be even more devastating than Chixulub-type events. About 250 million years ago, a large volcanic province developed in Russia called the Siberian traps. These multiple volcanoes emitted enough lava and solid ejecta to cover an area of over 1.5 million square miles (3.9 million km^2), about half the land area of the continental United States. It also emitted so much CO_2 that it warmed the climate and the oceans and changed their chemistry causing the extinction of 96 percent of marine life and 75 percent of land life.[3] This was the largest extinction event in earth history, even larger than Chixulub. Fortunately, it is the only recorded time an event of this magnitude ever happened.

Although not as catastrophic as asteroid impacts or volcanic provinces, single volcanic eruptions can still cause global devastation. Volcanoes capable of producing such eruptions have been termed "supervolcanoes" although this is a Hollywood term rather than a scientific term. These eruptions can cause a near nuclear winter called "volcanic winter" for a few years to decades, which also wreaks havoc on all ecological systems.[4] Yellowstone is the single supervolcano in the continental United States. Although it is best known for its geysers and hot springs in the national park, huge reservoirs of magma currently occur at depth beneath it, which could erupt in a supereruption, possibly at any time. The last such supereruption from this huge caldera or volcanic eruption crater was about 630,000 years ago. In this eruption, the volcano ejected about 240 cubic miles (1,000 km^3) of rhyolitic lava and ejecta, primarily ash. This ash was 39 inches (1 m) thick for 180 miles (300 km) in all directions around the volcano, but a layer covered the entire continental United States and most of Canada.[5] In comparison, the infamous 1980 eruption of Mount St. Helens in the Washington Cascade Range, which held the attention of the country and most of the world, produced about 0.24 cubic miles (1 km^3). It was 1,000 times as small as Yellowstone and yet powerful and deadly.

The Yellowstone eruption was not the only, most recent, or largest of the known supereruptions. The supervolcano Toba on the island of Sumatra in Indonesia had a major eruption about 73,500 years ago that dwarfed the Yellowstone eruption. The estimated output of this eruption is 700 cubic miles (2,800 km^3) of lava and ash. The huge amount of ash is estimated to have encircled the earth and filtered the sun for as many as six years, reducing the global temperature by 4–10°C locally. This cooling wreaked havoc on all life on the planet.[6] The event has been suggested to have caused a bottleneck in human existence where the species almost went extinct.[7] The theory, however, is controversial. Modern humans are certainly fortunate that there have been no supereruptions in historical times, or society would have collapsed. A worst-case scenario for a supereruption could be the demise of humans. This chapter focuses on the largest and most impactful volcanic eruptions in historic times with the understanding that they are nowhere near as large as the Siberian traps volcanic province or the supereruptions. If either of these develops in the future, there is nothing that anyone can do to prevent the catastrophe they will produce.

Volcanic Explosivity Index

In order to compare the intensity of volcanic eruptions, the volcanic explosivity index or VEI was developed.[8] The VEI of an eruption ranges from a low of 0 to a maximum of 8 based on a number of quantitative factors. The three main factors are (1) volume of ejecta or material emitted during an eruption, (2) height of the eruption column shot vertically above the volcano, and (3) duration of the eruption. Magma compositions and volcano types are commonly associated with VEI ranges. Basalt is from fissure eruptions and shield volcanoes, and they typically erupt with a VEI of 0 or 1. These types are called "Icelandic" and "Hawaiian" eruptions (VEI = 0–1) because they commonly occur in those areas. Andesites and dacites erupt from stratovolcanoes or cinder cones over a range of VEI up to 8 with eruption styles from the most intense "Pelean" eruptions (4–7) to moderate "Strombolian" (VEI = 1–3) and "Vulcanian" (VEI = 2–4) eruptions. They are named for volcanoes where the eruption types were observed. The intense "Vesuvian" eruption is also named for a volcano but equivalent "Plinian" eruptions (VEI = 3–7) are named for the Roman, Pliny the Younger who described this type of eruption at Mount Vesuvius in 79 CE. "Ultra-Plinian" is the most intense eruption (VEI = 5–8) for which Toba is an example.

Rhyolite volcanoes are the least common type. The thick and water-rich lavas can produce very powerful eruptions, up to VEI = 8, like Yellowstone. They produce lava domes and extensive ash deposits in many cases.

Thera Volcano: Eruption that Changed Ancient Western Civilization

The Greek island of Santorini is on most lists of top vacation destinations. This beautiful island has white buildings on steep cliffs overlooking the blue Aegean Sea. It is surprising that these crescent-shaped islands once encircled a huge volcano that is now just a 3 by 6-mile (5 by 10 km) oval crater or caldera. A small island called Nea Kameni in the middle of this crater still smolders and steams, forewarning of a possible return to activity. It is the beginning of the next volcanic cone. This now serene location frequented by tourists from around the world was the site of the natural disaster that most impacted the ancient history of western civilization.

Between the second and third millennia BCE, the eastern Mediterranean Sea region underwent a dramatic advancement in civilization. The Minoan civilization was arguably the most advanced of those in the region at the time. The Minoans were seafarers and centered on Crete and the surrounding islands. These people developed advanced culture, technology, written language, and art. This technology included indoor plumbing, double-hulled ships, and earthquake-resistant buildings, among many others. They dominated the entire Mediterranean Sea and the region at the time. There were other Minoan settlements around the region that shared in the great advances in civilization. One of these settlements was on the island of Thera. Current and ongoing excavations of buildings and other archaeological finds that are preserved under the thick ash deposits show that Thera was at least as advanced as Crete. In the excavated major city of Akrotiri on Santorini, multistory houses decorated with ornate frescoes, full of ceramics, and other artifacts demonstrate the interactions with the Minoan settlements on Crete. The problem with Thera is that it was located on a volcanic island with an active stratovolcano.[9]

Thera was just one of a number of volcanic islands in the Aegean Sea. They are the result of a small subduction zone within the complex plate tectonic interactions of the area. As with most subduction zones, the volcanoes are primarily basalt to andesite stratovolcanoes that are capable of explosive eruptions. The Thera stratovolcano appears to have erupted on a regular basis but not intensely enough to scare away the residents, thereby allowing advancement of the settlement. However, the intensity of eruptions increased considerably, causing the residents to evacuate the island, probably to Crete. Excavations of the island to date have uncovered no human skeletons in the ash of the final eruption, indicating that it was abandoned by then.[10]

The scientific evidence indicates that the Thera volcano began generating dangerous pyroclastic flows that deposited volcanic debris on the slopes in at least three layers but continued into the sea where they generated tsunamis. These waves struck shorelines from Turkey to Crete, and several were large. They damaged the shoreline communities and are recorded in archaeological evidence.[11]

The eruptions continued to intensify until the volcano underwent a catastrophic caldera collapse, which allowed seawater to pour into the hot interior of the volcano, producing a phreatomagmatic explosion. This explosion is interpreted to have been an ultra-Plinian type eruption with a VEI of 7, and it produced an eruption column estimated to have reached 19–22 miles (30–35 km) in altitude, which is within the stratosphere. This eruption has been compared to the 1883 Krakatoa volcanic eruption in Indonesia. That explosion was heard as far as 3,000 miles (5,000 km) away. If the comparison is accurate, the reverberation from the Thera explosion was probably heard from the British Isles to Persia. The explosion also produced a huge tsunami estimated by some to have reached 115–492 feet (35–150 m) high. Fresh sediments were deposited on the top of an island near the volcano at an elevation of 230 feet (70 m). This wave devastated the northern coastline of Crete, 68 miles (110 km) to the south. The huge wave spread out around the whole Mediterranean basin with heights of 15–28 feet (5–8 m) in Turkey and much of the Middle East. On Crete, inundation distances were calculated to be 820–1,475 feet (250–450 m) inland with waves as high as 92 feet (28 m) locally.[12,13]

The waves were so large that they scraped up already deposited sediments from the seafloor and swept them into a chaotic layer of mixed sediments that has been termed "homogenite." These deposits are up to 65 feet (20 m) thick, where they represent slope failures of sediments generated by the tsunami passage. Otherwise, the deposit is up to 9 feet (3 m) thick locally and occurs throughout the Mediterranean basin.

The exact timing of the eruption and resulting tsunami is somewhat problematic because of problems with background Carbon-14 ratios at the time. Ages from Carbon-14 dating extend from 1500 to more than 1650 BCE. Spikes in sulfur deposition in Greenland ice cores show an extended period from 1650 BCE to the mid-1500s; however, an eruption in Alaska in 1628 BCE complicates these data. New tree-ring analyses from the region indicate a probable age between 1560 and 1600 BCE.[14]

The eruption of Thera and the resulting tsunami are credited with extensive societal impacts around the region and even changes to Western civilization. For certain, the tsunami decimated the Minoan civilization on Crete. This weakened them so much that it ultimately led to the downfall of the Minoans. Their weakening led to the emergence of the Myceneans, the inland Greeks, and the replacement of the dominant language and writing. The emergence of the Myceneans led to the Golden Age of Greece, which was a major factor in the shaping of Western civilization. The story of the rise and downfall of Thera is interpreted by some to have really been Plato's legend of Atlantis. This interpretation, however, remains controversial.[15]

Even more intriguing, some have interpreted the sequence of volcanic and tsunami events at Thera to reflect the story of the Exodus of Israelites from Egypt. Many of the ten plagues can certainly be explained by the processes of an active volcano.[16] Stratovolcanoes emit huge amounts of ash on a regular basis. If ash fills the sky and is thick enough, the reflection of the filtered sun can cause all liquid to appear blood red. If the ash gets thick enough, it can blot out the sun completely, leaving an area in blackness. When the ash settles on the ground and surface water, sulfur in the ash can acidify surface water, killing all of the aquatic life and encouraging insects to infest their rotting corpses. A good example of this was the 1912 eruption of Novarupta in Alaska, the largest eruption of the twentieth century. So much sulfur leached from the ash into the streams that the local salmon population was decimated and did not recover their populations for a decade. The rotting fish in the streams became breeding grounds for so many flies that it became impossible to visit the area.[17]

Finally, the famous legend of Moses parting the Red Sea could be the result of the Thera eruption. When a tsunami approaches the shoreline, the sea retreats substantially for several minutes. This could seem like control over the sea. It also leaves time for people to walk in an area that is usually underwater. After this, the tsunami strikes the shore with such force that it could destroy even the largest of armies. These explanations of biblical events using the Thera disaster have been suggested, but the details and timing of them have yet to be proven.[18]

If a similar powerful eruption of a volcano in the Aegean Sea occurred today, it would cause the worst natural disaster in history.

536 CE: Start of the Dark Ages

There are some cases where terrible things happened and are well-known in history that turn out to be the result of a natural disaster or perhaps several of them. This is the case for the well-known Dark Ages, although that term is out of favor in the historical community.[19] This period extended from 536 to 680 CE and was marked by wars, pestilence, famines, shifts in power, a nail in the coffin of the great Roman Empire, and the complete regression of human cultural and technological advances in some cases. Appropriate or not, these were truly dark ages for many, and they resulted from several volcanic eruptions.

This period began abruptly in 536 CE. Dr. Michael McCormick, the medieval scholar, stated that 536 CE "was the beginning of one of the worst periods to be alive, if not the worst year" ever.[20] Certainly, there have been many other terrible years and periods, so this statement is debatable, but the intensity with which this period emerged is frightening. In March of 536 CE, a low-hanging dry haze abruptly appeared, heralding the "days of darkness" and accompanying strange weather patterns which were reported in historical journals. It is claimed that this haze plunged Europe, the Middle East, and parts of Asia into a period of weak sunlight for as much as eighteen months and was accompanied by a sharp drop in temperatures. This began the Late Antique Little Ice Age, as it is known, which lasted for more than 120 years, from 536 to 660 CE.[21,22]

This cooling is well demonstrated in tree-ring and chemical studies (carbon isotope ratios), as well as historical records. In the Summer of 536 CE, temperatures were up to 4.5 °F (2.5 °C) below normal in Europe. There are records of snowfall in China and Europe during that summer, accompanied by droughts and resulting crop failures in some areas and great floods in others. The cool weather continued into the following winter, marking the most severe cooling in the Northern Hemisphere in 2,300 years.[22,23] However, this was not the only cooling event in this period of history. There were subsequent markedly cold periods in 540–541 CE and in 545–546 CE. Just as society was coming out of one cold period, another would strike, prolonging the pain and suffering.[24]

As a result of this cooling, drought, and other climate impacts, society went into chaos. Droughts and famines in the steppe region of Asia caused the Mongol hordes to venture west and invade eastern Europe and the Middle East. A major drought in India led to unrest and, ultimately, the overthrow of the Gupta Empire. The harsh conditions led to the collapse of the Sassanid Empire as well. The Byzantine Empire defeated the weakened Vandal Kingdom in Africa. In Scandinavia, the cold, drought, and famine compelled Viking leaders to dump gold into ponds and streams to appease the gods. The conditions are thought to be the inspiration for the Ragnarok legend about the death of the gods through natural disasters. There are reports of famines in Ireland at this time as well. The climate changes also led to the demise of cultures in South America including the fall of Teotihuacán.[25] There are even theories that the climate impacts play a part in the King Arthur legend in England.[26]

However, the biggest changes in history during this time involved the remains of the Western Roman Empire and the area around the Eastern Empire. During this time, they were engaged in war with the Goths on the Italian peninsula, later named the "Gothic War." Although they were victorious, the war lasted twenty years and consumed immense resources of the empire. Droughts and famines caused unrest along many of the other Byzantine borders. They were barely holding the empire together when an even worse tragedy struck. The cold, wet conditions encouraged vermin and pestilence and starting in the Roman port of Pelusium, northern Egypt, in 541 CE, a horrible bubonic plague took hold. It swept rapidly through the eastern Roman Empire and killed about 35–55 percent of the population from 541 to 549 CE. Even though Emperor Justinian I had done an outstanding job, holding the empire together and even advancing it through this difficult period, he also contracted the disease. After that, this plague was termed the Plague of Justinian. It is the second most deadly of the plagues.[27]

The empire contracted and had a hard time holding together at all. They were invaded from many sides. All of the lands in Italy were lost. The extensive trade ship routes developed during the first century CE were nearly eliminated by the mid-sixth to early seventh centuries. Italy reverted to a near-prehistoric existence like that which prevailed from before the time of the Etruscans.

However, life was not terrible for everyone. The lower temperatures in Arabia made life much more tolerable, and civilization advanced. Some previously arid areas became able to sustain livestock and agriculture. This increase in fertility led to a sizeable increase in the population over a few generations. This increased population increased the influence and power of Arabia. The religion of Islam was founded in this area during this time, which helped unite the newly enlarged population. With this focus and manpower, Arabia challenged the Byzantines and Persians. By the end of this dark period, Muslim armies had conquered Palestine and Syria.[28]

All of this tumultuous history has been known for many centuries. However, the reason for the huge climatic shift was not known. It took extensive scientific research to provide the evidence that the cooling was global and its probable cause. The sharp cooling is documented in Europe, the Alps, Scandinavia, and Russia using tree-ring data, and they show three events. The evidence that the events were volcanic in origin lies in large spikes in sulfates in ice cores taken from the Greenland ice sheets at both 536 and 540 CE. The 546 CE spike is less certain. These spikes or high concentrations of sulfates in the ice result from high amounts of sulfur released during volcanic eruptions. Surface chemistry in unusual elements and carbon isotopes also supports a volcanic origin for the cooling. This makes the three cooling events "volcanic winters."

The problem has been to identify the massive volcanic eruptions that caused the cooling events. This is difficult because many of the suspect volcanoes are in isolated locations with no historical records. The researchers have suspected several different volcanoes as the culprits. At first, the suspect was a proto-Krakatoa volcano, but Mount Rabaul in New Britain was also suspected. At present, the 536 CE cooling is suspected to have resulted from an eruption from a volcano in Iceland, possibly Krafla. The 540 CE eruption is better agreed to be Ilopango in El Salvador. Between a northern and equatorial eruption, the entire northern hemisphere would have been blanketed in ash.

Modeling exercises were conducted on the possible eruptions by researchers. They predict a VEI of 6 that ultimately includes caldera collapse and flooding of seawater into the volcano. A resulting phreatomagmatic explosion could generate a plume 16 to >32 miles (25 to >50 km) high that could inject 12–24 cubic miles (50–100 km^3) of sulfate aerosols into the stratosphere. Additionally, it would have injected as much as 18 cubic miles (75 km^3) of fine ash. A volcanic eruption of this volume could have caused global cooling by as much as 5.4 °F (3 °C) for up to the next five years.[29]

Laki Fissure Eruptions: Poison Gas Emissions

In addition to the danger of the explosions, heat from lava and pyroclastics, and climate change from ash and aerosols, chemicals from volcanic eruptions can actually poison people and the environment over a large area. Although the infamous Laki eruption, Iceland in 1783 is known for its impact on the climate, and its poisoning of the environment is really what changed history.

The Laki or Lakagígar volcano began erupting on June 8, 1783. It erupted from a fissure or crack that had 130 craters along its length of 16.8 miles (27 km). This eruption persisted for eight months until February 7, 1784, when it finally quieted. The eruption is also known as the Skaftáreldar ("Skaftá fires") or Síðueldur, and it was the single largest outpouring of lava in historical times. Some 3.5 cubic miles (14.7 km^3) of basalt lava poured from the vents, and about 0.2 cubic miles (0.91 km^3) of solid pyroclastic materials were also emitted. This was the largest pyroclastic emission from any volcano in Iceland in more than 250 years.[30]

During the first five months of this eruption, Laki experienced ten eruptive pulses, each with an explosive phase lasting about twelve hours to four days, and they were followed by a longer period of lava fountaining. Early activity was explosive and included violent Strombolian to sub-Plinian eruptions. The lava fountains are estimated to have reached approximately 2,600–4,600 feet (800–1,400 m) in height. The peak discharge of lava is estimated to have ranged between 176,573 and 211,888 cubic feet (5,000–6,600 m^3) per second, but there are estimates of up to an impressive 295,000 cubic feet (8,600 m^3) per second. In comparison, the average discharge of the Mississippi River into the Gulf of Mexico is 593,003 cubic feet (16,792 m^3) of water per second. The eruption was most active during the first forty-five days, and it was followed by a steady decline over the next three months. By late July and the end of the fifth pulse of the eruption, about 60 percent of the lava had been emitted. By October and the end of the tenth pulse, some 93 percent of the lava had been erupted. In the final 3.5 months, the eruption was characterized by a quiet release of lava and gas.

The massive lava release covered 218 square miles (565 km^2) across Iceland. The lava flow extended about 15 miles (24 km) southwest from the volcano and downhill toward the ocean. The lava flow is 10 miles (16 km) at its widest section, and it has an average thickness of about 100 feet (30 m). The lava filled two large river valleys where the water had been dammed by the lava, and the thickness reached several hundred feet.

However, the real impact of Laki resulted from the emission of aerosols into the atmosphere. The eruption columns are calculated to have attained altitudes up to 8.1 miles (13 km) during the earliest phases of the eruption but remained above 6.2 miles (10 km) high for the first three months. Although quite high, these elevations are much lower than eruption columns from large stratovolcano eruptions and they more readily impacted the earth's surface. However, these eruption columns reached the elevation of the westerly jet stream, and emissions were spread throughout Europe and areas beyond. In addition to ash, Laki also injected about 8 million tons (7.3 million mt) of hydrogen fluoride and around 132 million tons (120 Tg) of sulfur dioxide

(SO_2) into the upper troposphere and lower stratosphere. Over the first three months of the eruption, SO_2 concentrations exceeded 20 ppb locally, and sulfuric acid was about 2 ppb compared to preindustrial values of 0.1 ppb of SO_2. The sulfur dioxide was deposited onto the surface by fallout or converted into sulfuric acid in the atmosphere and then produced acid rain and fog.[31]

The worst effects of the acid fallout and precipitation were within Iceland where a dry fog known as the "Laki haze" enveloped the country for at least five months. Fallout of sulfur and fluorine compounds from this fog poisoned the grazing areas. The result was that some 80 percent of the sheep, 50 percent of the cattle, and 50 percent of the horses in Iceland perished. By the end of 1785, between 20 and 25 percent of the human population, more than 10,000 people, also died from acid poisoning or in the ensuing famine.[32] Ash and sulfur fallout drifted into Great Britain, causing the summer of 1783 to be known as the "sand summer" from all of the falling ash as a result. The Laki haze also spread across Europe. By June 26, the fog reached Lisbon, Portugal, and St. Petersburg, Russia, and it reached Moscow by June 30. It was over Tripoli by the end of June, and by July, it had spread to the Altai Mountains in central Asia, 4,350 miles (7,000 km) from Laki. Later, the Laki haze had spread across the entire Northern Hemisphere from 35° north latitude to the North Pole. There are even reports of acid fallout from Laki reaching Brazil.[33]

The sulfuric acid fallout from the haze devastated vegetation and human health all across Europe. In Iceland, acid rain left sores on animals and humans and burned holes in the leaves of all exposed vegetation. Crops failed in Iceland but also had low yields or failed throughout Europe that summer and for several years thereafter. There were fish kills in Scotland caused by acid surface water from the fallout. Even livestock suffered and perished all across western Europe, east to Italy, for years. Inhaling the haze caused shortness of breath, headaches, fatigue, and palpitations of the heart in people. It was especially hard on the very old, the very young, and those with respiratory ailments. England reported extremely high mortality rates in the summer of

1783, and there were increased respiratory disorders and mortality rates all across Europe.[34]

As stated, the ash and aerosol from the Laki eruption also affected the global climate at least initially. After a warm July, the remainder of 1783 was unusually cold in Europe and most other areas. The Nile River watershed suffered a drought, producing a record low river level. The Altai Mountains suffered harsh frosts during August. A severe drought also struck India through the Yangtze River region in China, and the summer was cold all over China. The excessive precipitation and low temperatures resulted in widespread rice crop failure in Japan, causing the most severe famine in its history.

The subsequent winter (1783–4) was among the most severe in North America and Europe in 250 years, dropping temperatures by an average of 5.4 °F (3 °C) below normal. The winter included long periods of freezing temperatures. There was extreme weather in Great Britain, France, and Scandinavia with excessive cold and snow. In January, the water around the Danish islands was frozen over, and many other rivers were frozen. In North America, New England suffered a record period of subzero temperatures, and New Jersey had record snow accumulations. Ice caused a record closure of the harbors and channels in the Chesapeake Bay of Maryland and Virginia. Even the Mississippi River was filled with ice rafts at New Orleans. This cooling, however, was short-lived as the temperatures in Europe were actually above average over the following year, although droughts and famine in Asia persisted for another year or two.[35]

Between the heavy acid fallout and unusual weather, at least initially, there was widespread famine and death from the health impacts and starvation. At least 23,000 people perished in Great Britain and more than 16,000 died in France in 1783–4 alone. The famine across Egypt, China, and Japan caused tens of thousands of deaths. It is estimated that up to 6 million people died during the entire event. Fortunately, a large-magnitude Laki-type event only occurs approximately every 500 years on average. The disruption of the food supply from Laki sulfur poisoning caused the economy of Western Europe to falter for several years. The lack of food led to widespread hunger, poverty, and unrest. The lack of food and weak economy persisted until 1788 and is credited as a major contributing factor to the French Revolution in 1789. The phrase credited to Marie Antoinette "let them eat cake" was in response to the shortage of bread for the public, which was at least partly caused by Laki.[36]

Protection Against Volcanic Eruptions

Volcanic eruptions are natural processes that happen on a daily basis, and there is no way to control them. Certainly, there is no way to stop a supereruption or to even plan for it. Eruptions that poison areas or change the climate are also impossible to plan for, although global trade can redistribute food and other necessities enough to at least partially alleviate the hardships caused by them. Volcanic hazards can also generally not be stopped, but with good prediction methods and evacuation plans, people can be saved. They could have been used in Thera and in many other destructive eruptions as well. For example, the famous eruption of Mount Vesuvius in 79 CE produced a huge summit eruption but also extensive pyroclastic flows that destroyed the town of Herculaneum and entombed the town of Pompeii. It also killed an estimated 20,000 people. Now, there are more than 3.1 million people in the greater Naples, Italy area alone, all in the path of the next eruption.[37]

That eruption cannot be stopped but if the volcano is carefully monitored and the population is properly educated on evacuation routes and procedures, then a second disaster might be averted. Monitoring of volcanoes takes on several forms. When magma moves toward and into the volcano, it forces the surrounding rock to crack, producing earthquakes. Some of these earthquakes that are specific to volcanoes produce distinctive screw-shaped patterns on seismographs and, as a result, are called "tornillos" or small screws. Active volcanoes can be seismically monitored on a continuous basis. When magma begins to move into the volcano, it causes the cone to inflate and bulge. Even the ground around the volcano can be uplifted. Elevation on and around the volcano and slope alignment of the cone can be carefully monitored including using long-distance laser monitors, which detect even millimeter-scale changes. These changes can signal an impending eruption. Gases escaping from the volcanic vent can also change their chemistry before an eruption. Monitoring of this gas chemistry is another method to determine if an eruption is imminent.

Prediction of an eruption must be coupled with education, public awareness, and preparation of emergency services and public transportation services. Once the eruption is even suspected, the whole area must go into alert. Typically, there are different levels of preparedness depending on the level of activity. These alerts are advanced to higher levels as the eruption becomes more imminent. The residents are educated on the meaning of the alerts, preparation for evacuation, and evacuation procedures, if needed. The government and emergency services are also educated on the procedures and timing of evacuation. If the monitoring, alert, and evacuation processes are carried out properly, residents can be relatively safe in even highly destructive eruptions.

1994 Rabaul Eruption: Papua New Guinea

The active volcano on Rabaul, an island in Papua New Guinea was quiet for more than fifty-one years. Over that period, the population of Rabaul town at the foot of the volcano had increased to about 50,000 people. As of September 15, 1994, there was no indication of an impending eruption as a leveling survey of ground elevation was the same as it had been for years. The situation changed quickly at about 3:00 a.m. on September 18 when two large earthquakes occurred in the volcano caldera, one with a strong magnitude of 5.1. This began twenty-seven straight hours of vigorous seismicity.

This abrupt seismicity alarmed the local officials and they issued a voluntary evacuation alert by the afternoon. Residents began evacuating by ferry to the main island of Papua New Guinea. Officials contacted the national volcanology survey and declared a Stage 2 alert by 6:00 p.m. on September 18. This stage means that an eruption is expected within weeks to months and evacuation activity increased. By midnight, the Provincial Disaster Committee was informed that the eruption was imminent. Residents had already begun an orderly evacuation and were taking the situation seriously. Within 2 hours, institution of a Stage 3 alert was recommended. This meant that an eruption of Rabaul was expected within days to weeks. However, evacuation was going so well that it was decided not to announce it for fear of causing panic. By 7:00 a.m. on September 19, the town was deserted with all residents having been evacuated.[38,39]

Meanwhile, the eruption sequence continued. By 6:00 a.m. on September 19, offshore areas had been uplifted enough to be out of the water. In one area, a tide gauge was elevated by almost 20 feet (6 m). On the west and south coasts of nearby Matupit Island, the shoreline was shifted 230 feet (70 m) to the south. By 6:03 a.m., smoke began rising out of the volcano and within minutes, a large ash cloud emerged. Large ejecta were shot out of the caldera and landed in the water up to 0.6 miles (1 km) from the volcano. The volcano erupted from two vents at once, Vulcan and Tavurvur. A strong Plinian eruption proceeded with an eruption column about 12 miles (20 km) high. Large pyroclastic flows covered the slopes of the volcano and they generated tsunamis upon impacting the surrounding waters, some with waves as high as 16 feet (5 m). The eruption continued until October 18 with varying levels of activity. This eruption has been assigned a VEI of 4 which is quite strong. However, a natural disaster of catastrophic proportions was averted thanks to modern monitoring and emergency management.

1991 Mount Pinatubo Eruption: Philippines

Another even more impressive example of predicting an eruption using monitoring technology and evacuating an area involved the 1991 eruption of Mount Pinatubo in the Philippines. The reason that it was so impressive is that Mount Pinatubo was the second-largest volcanic eruption of the twentieth century with a VEI of 6. Further, the population within the area impacted or potentially impacted by the eruption was as much as 6 million people. In spite of this recipe for a catastrophe, only a handful of people died in the eruption. This would have been the best example of a tragedy avoided except that a different natural disaster occurred the next day, a typhoon, that killed a large number of people. That tragedy soured a potential victory.

The Philippines is no stranger to natural disasters. It lies in the main track of Pacific typhoons, but it also hosts a convergent margin that produces volcanoes and major earthquakes. The events that led up to the Mount Pinatubo eruption began on July 16, 1990, with a very strong magnitude 7.8 earthquake about 60 miles (100 km) northeast of the long inactive Mount Pinatubo on Luzon, the Philippines. The earthquake generated a landslide on Mount Pinatubo, and there was an increase in steam venting in a nearby geothermal area. In March and April 1991, magma began rising from 20 miles (32 km) below the volcano, accompanied by small earthquakes and causing steam releases and explosions. The small earthquakes continued through May and into early June, and sulfur gas was emitted from the summit of the volcano.[40, 41]

The first eruptions of Mount Pinatubo began on June 7–12, 1991. Initially, a lava dome formed in the caldera. However, on June 12, an explosive eruption occurred. This was just the start of the increased activity. On June 15, 1991, an explosive Plinian to ultraPlinian eruption with a VEI of 6 took place. It produced an ash cloud 28 miles (40 km) high, emitting more than 2.4 cubic miles (10 km^3) of ejecta. The ash and larger lapilli or pebble-sized pumice fragments, however, were dispersed throughout the area by an incoming typhoon. Ash was deposited all the way to the Indian Ocean, and the ash cloud was tracked by satellites several times around the earth. Extensive pyroclastic flows then descended the slopes of Mount Pinatubo at 60 miles per hour (100 kph) and filled the valleys surrounding the volcano with deposits up to 660 feet (200 m) thick. As the eruption ended, the summit of the volcano collapsed, leaving a 1.6-mile (2.5 km) wide crater-like caldera.

The eruption of Mount Pinatubo could have had a catastrophic death toll. Fortunately, the US military maintained the large Clark Air Force base on the edge of Mount Pinatubo and the Subic Naval base on the edge of the eruption area. These two bases had significant numbers of military personnel and very expensive weapons. As a result, the military very carefully monitored the eruption. When the eruption began in earnest, the military removed all personnel and weaponry, but they also warned the 200,000 nearest residents to evacuate the area, and more than 60,000 people complied. This action saved at least 5,000 lives, but more likely tens of thousands of lives. The eruption caused an estimated $700 million in damage.

This could have been a complete victory over a natural disaster, but unfortunately, it was not. Typhoon Yunya struck the eruption impact area the next day, and torrential rain soaked the very hot, freshly deposited ash. The resulting boiling mud created a lahar or volcanic mudflow that flooded beyond the evacuated area, killing 847 people. This is a problem with volcanic ash deposits. After more than five years, the thick ash deposits retained most of their original heat. They were still as hot as 900 °F (500 °C) by 1996. Surface and ground water still exploded in contact with ash. Ash deposits were also still being remobilized as lahars during heavy rainstorms for years.

Decade Volcanoes

Active volcanoes in populated areas are especially prone to causing natural disasters. In response, the scholarly organization, International Association of Volcanology and Chemistry of the Earth's Interior (IAVCEI) of the International Decade for Natural Disaster Reduction which were in the 1990s, designated the "Decade Volcanoes." Designation of the Decade Volcanoes was designed to bring attention to several active volcanoes in populated areas to enhance scientific research, public awareness, understanding, and literacy of volcanoes and the hazards they pose. Sixteen volcanoes were designated as Decade Volcanoes.[42] They are

1) Avachinsky-Koryaksky, Kamchatka
2) Colima Volcano, Mexico
3) Mount Etna, Italy
4) Galeras Volcano, Columbia
5) Mauna Loa, Hawaii
6) Merapi Volcano, Indonesia
7) Niragongo Volcano, Republic of Congo
8) Mount Ranier, Washington
9) Sakurajima Volcano, Japan
10) Santa Maria/Santiaguito Volcano, Guatemala
11) Santorini Volcano, Greece
12) Taal Volcano, Philippines
13) Teide Volcano, Canary Islands
14) Ulawun Volcano, Papua-New Guinea
15) Unzen Volcano, Japan
16) Mount Vesuvius, Italy

These volcanoes are active and have the potential to produce powerful eruptions. They are close to populated areas and also have the potential to produce catastrophic natural disasters. For this reason, the Decade Volcanoes are carefully monitored. They have sophisticated monitoring systems including gas analysis, seismic monitoring, and level monitoring. The residents of these areas are educated about the signs of potential volcanic eruptions, alert levels, and evacuation routes and procedures.

Eldfell Eruption, Iceland: Controlling a Lava Flow

The only option with most volcanic hazards is to get out of the way as far as possible. However, there was one case where a highly destructive lava flow was prevented from destroying a town, although no human lives were at risk. The Eldfell volcano formed on the island of Heimaey off the southern coast of Iceland. Iceland is a volcanic island situated on the divergent boundary of the Mid-Atlantic Ridge. Most of the numerous eruptions are not explosive or even particularly threatening to human settlements. The Eldfell eruption of 1973 was almost an exception.

On January 21, 1973, eruption precursors began about 10:00 p.m. with a swarm of earthquakes around Heimaey. The frequency of earthquakes increased to 50 per hour between 1:00 and 3:00 a.m. on January 22. The earthquakes then subsided until they stopped at 11:00 a.m. From 11:00 p.m. to 1:34 a.m. on January 23, stronger but shallower earthquakes occurred. At 1:55 a.m., a fissure 985 feet (300 m) long developed on the east side of the island near the center of the town of Heimaey. The fissure grew to 1.25 miles (2 km) long, crossing the entire island. A spectacular fissure eruption of VEI = 0–1 developed with more than forty lava fountains 164–492 feet (50–150 m) high along the entire length. The fissure then lengthened to 1.9 miles (3 km). The main eruption concentrated to a single 0.5-mile (0.8 km) vent that emitted lava and ejecta at about 3,531 cubic feet (100 m^3) per second. It became a Strombolian eruption of VEI between 1 and 2+ that built a cone about 328 feet

(100 m) high. By February 15, this cone had grown to 722 feet (220 m) high. The eruption column during this time reached a maximum of 5.6 miles (9 km) high.[43,44]

Eldfell was a relatively powerful eruption for a basalt volcano but not an explosion threat for Heimaey. However, the thick lava flow spread east, northeast, and north and ash up to 16.4 feet (5 m) thick covered the town. The lava flow was up to 328 feet (100 m) thick in some places and burned and crushed every tree and building in its path.

Fishing was and is the main industry for Heimaey. A storm before the eruption forced the island's fishing fleet of seventy ships into the harbor. When the eruption began, the Icelandic Civil Defense Organization sounded all of the sirens and the residents boarded the fishing boats for evacuation. The evacuation began at 2:30 a.m., a half hour after the eruption began, and many of the residents were also flown out from the small airport. The civilian population of 5,300 people was evacuated within six hours. Only 250 emergency personnel and volunteer workers remained to clear the ash from houses. At least, 400 houses collapsed or burned. By February, the ash emissions abated, and the remaining homes were relatively safe.

The lava extrusion declined from 3,531 cubic feet (100 m^3) per second to 2,119 cubic feet (60 m^3) per second by February 8 and 353 cubic feet (10 m^3) per second by mid-March. However, so much lava had been extruded that large lava flows were heading toward town and the harbor. A submarine lava flow cut the electric and water supply to the island. Then the main lava flow encroached on the harbor. This was a serious problem because if it cut off the harbor, it would destroy the fishing industry, which provided one quarter of the annual catch for all of Iceland. Fishing was the only source of income for the town, so it would end the town if fishing were eliminated.[44,45]

The Iceland government attempted an experimental procedure of spraying seawater onto the front of the lava flow in an effort to harden it and stop its forward movement. This had been attempted in Hawaii and on Mount Etna, Italy, on a much smaller scale and with minimal success. The procedure was tested on February 6 with seawater pumped at 26.4 gallons (100 L) per second on the front of a lava flow. The sprayed water boiled to steam on contact. It took a while but the lava hardened on the front and the flow slowed significantly. In response, the operations were greatly expanded, and water was sprayed more strategically on the lava flows.

On March 1, workers brought in a dredging boat that pumped 106 gallons (400 L) per second to attempt to slow a rapidly advancing lava flow. They placed a network of pipes on the lava that sprayed seawater over an area of 129,167 square feet (12,000 m^2) at a time up to 427 feet (130 m) back onto the flow. However, the supports fell, melted, or caught on fire; there were also toxic CO gas fumes emitted from the lava.

Despite these efforts, about one-fifth of the town was engulfed in lava by the end of March, and the lava flow was continuing to overtake it. The United States had sent thirty-two oil pumps with pumping capacities of 264 gallons (1,000 L) per second each. They significantly increased the volume of seawater being sprayed onto the lava, and the flow slowed dramatically and finally stopped with the fishing harbor saved and still accessible. The eruption continued until the middle of April at a rate of 177 cubic feet (5 m^3) per second, but no more problems were encountered.

The Eldfell lava flows created at least 1 square mile (2.5 km^2) of new land on the island. The volcanic ejecta was used as fill material to extend the runway at the airport and as foundations for 200 new houses. The heat from the lava flows has been used to power several 40 megawatt steam turbine generators, providing electricity to the island, as well as hot water for home and industrial usage. The new land now acts as a breakwater for the harbor, improving its quality. More than 80 percent of the of the displaced population had returned by 1975.

The Takeaway

For single-event natural disasters, volcanic eruptions are second only to large asteroid impacts with the earth in terms of their ability to change human and all other history. Fortunately, they are so rare that they are not much of a concern on a daily basis. However, they can at least change human history if not wipe out the species. Supereruptions of supervolcanoes are a close second to impacts in terms of damage to human society if not the species. Although they are much more common than large impacts, they are still rare enough that most people do not worry about them. However, there is nothing humans can do about either of these, and history will never be the same if either occurs.

For the smaller, more common eruptions, in addition to the direct hazards like lava flows, pyroclastic flows, volcanic, or phreatomagmatic explosions, tsunamis, and volcanic bombs, short-term climate change and poisoning can cause great changes to human history. Besides lava flows, which humans have had limited success in controlling, there is no way to control any of the other hazards at this point in history. Fortunately, monitoring technology is sophisticated enough to alert and allow evacuation of human populations from the danger zone around the volcano for the direct volcanic hazards like pyroclastic flows, explosions, tsunamis, bombs, and lava flows in most cases. These efforts have already saved tens to hundreds of thousands of lives if not more.

Climate change and large-scale poisoning are another story. They cannot be predicted and nothing can be done about them even if they could be. Fortunately, global trade allows food to be redistributed on a short-term basis. This ability could make even significant climate changes and poisonings less impactful.

4

Earthquakes

Earthquakes are among the most deadly and destructive natural disasters. We can tell where almost all of them will occur, but there is no way to predict when they will occur at this point. At least 99 percent of earthquakes occur at plate boundaries with very few being classified as intraplate if they are not. Earthquakes occur on faults. Tectonic stress builds up on the rock in the fault until it breaks. This breakage is an earthquake, and the built-up stress is converted into the seismic waves that cause death and destruction. The part of the fault that breaks is the focus or hypocenter, and the epicenter is above it on the surface. The earthquake waves radiate outward in all directions from the focus like the rings of waves from throwing a rock into a pond. The rings on a pond are only on the water surface, but in earthquakes, some waves also travel throughout the volume of the earth.[1]

The two basic earthquake wave types are body waves and surface waves. Surface waves are like the rings on the pond radiating from the epicenter, but only on the surface. They are by far the most dangerous. The two types of surface waves are Rayleigh waves and Love waves. In contrast, body waves travel throughout the volume of the earth. The two types of body waves are P-waves and S-waves. P-waves are the first to arrive at the seismograph, and S-waves arrive second. All types of waves are detected on a seismograph, and the digital readouts are called seismograms, and the location and magnitude or power are determined from analyzing them.

Damaging earthquakes occur in clusters. A slip on a fault produces an earthquake, and stress is released on the area that slipped. This increases the stress on the fault adjacent to the slipped area. These areas then slip, producing additional earthquakes which increase the stress on other parts of the fault. The sequence of earthquakes constitute foreshocks, a main shock, and aftershocks, but they need to be assigned after some of the sequence is over. The main shock is the most powerful of the earthquakes. Once the main shock is identified, then earthquakes before it are foreshocks and the earthquakes after it are aftershocks. If there is only one earthquake, it is a main shock. This is common for smaller earthquakes. Larger earthquakes can be followed by aftershocks for a year or more. Early aftershocks can be strong and in rapid succession but they tend to get weaker and less frequent.

The earthquake magnitude and intensity are measures of its power and destructiveness, respectively. Charles Richter first defined the magnitude scale for earthquakes. It is primarily a log function of the wave amplitude meaning that each magnitude number is more than ten times as powerful as the next lower number. A magnitude 7 earthquake is more than 10 times as powerful as a 6 and more than 100 times as powerful as a 5. However, distance from the focus is also a factor as the wave amplitude decreases with distance. Earthquake magnitude was previously referred to as Richter magnitude, but now it is moment magnitude, which uses data from multiple seismographs to determine the true power.[1,2]

Basically, most earthquakes with a magnitude of less than 3 can only be felt at the epicenter, if at all. Magnitudes of 4 and 5 can be felt widely but rarely do major damage. Magnitudes of 6 do extensive damage and can be deadly with large numbers of casualties depending upon the area. Magnitudes of 7 and 8 are major earthquakes and always cause extreme damage and numerous deaths depending on the area. There have only been a handful of earthquakes with magnitudes of 9 but they are always catastrophic.

The intensity of an earthquake is a measure of the damage caused by shaking from surface waves. An intensity scale was developed by Mercalli based on observational damage, and it had ten levels but was later expanded to twelve levels. It was revised to the Modified Mercalli Scale, which is still in use today. The scale is in Roman numerals, and it is for a specific area for a specific earthquake. Intensity measurements from different areas or from different times cannot easily be compared. Assigning intensities requires a settlement with buildings and witnesses of the event. However, building codes and the quality vary by area. Poorly constructed buildings are more strongly damaged by seismic waves than well-constructed buildings. Therefore, an earthquake can impose intensity VII in one area but intensity X elsewhere. In addition, different types of rocks, sediments, and soils transmit waves at different strengths. Areas with wet sediments, soils, or fill amplify surface waves stronger, which results in worse damage. Distance from the epicenter also controls the amount of shaking.

The origin of more than 99 percent of earthquakes is movement on a fault. When a fault ruptures in one place, it no longer helps to support the extreme stress that has built up along the length of the fault. The stress elsewhere on the fault, but especially adjacent to the rupture, increases in stress. If the stress is strong enough, they break and generate earthquakes as well. This process can continue until the whole fault has slipped.

Faults are classified as those with mainly vertical movement, termed dip-slip faults, and those with lateral movement, termed strike-slip faults. In dip-slip faults, the fault is generally a sloping surface. The footwall is the rock under the fault slope, and the rock above is the hanging wall. The motion of the hanging wall relative to the footwall determines the kind of fault, and it is relative to gravity. In a normal fault, the hanging wall rockslides down the fault slope, which is normal with regard to gravity. Normal faults form in extension and are common in divergent plate margins. The hanging wall rock is pushed up the fault slope in reverse faults, which is the reverse of gravity. If the fault slope is 30° or less, reverse faults are thrust faults, and on plate margins, they are megathrusts. Reverse and thrust faults accommodate contraction. Strike-slip faults are very steep and are classified by the relative lateral movement on the fault. If a person looks across a strike-slip fault, the opposite side moves either left or right. If the opposite side moves to the left, the fault is called a left-lateral or sinistral strike-slip fault. If it moves right, the fault is called right-lateral or dextral strike-slip.

464 BCE Sparta, Greece Earthquake

An earthquake struck the major city-state of Sparta in 464 BCE that changed the course of Greek history. It was likely the first well-documented earthquake in recorded history being chronicled in the writings of Thucydides, Diodoros, Aelianos, and Plutarch, among others.[3] It occurred in the right place and at the right time in Greek history that its effects would help dictate the actions of the next century and help set the stage for an even longer period. Ultimately, it would help shape the Greek influence on Western civilization.

Greece is a tectonically active area with both regular earthquakes and active volcanoes. The southern islands sit on the Hellenic Arc, which is a small subduction zone in the eastern Mediterranean Sea. It accommodates convergence between Africa and Europe with African ocean crust being consumed beneath Greece. The extrusion of Turkey westward into the Mediterranean further complicates the geology and increases the seismicity.[4] Italy and Greece are, by far, the most active and dangerous countries in Europe in terms of natural disasters.

Just prior to the earthquake was a historically important time for the area. The first war between the Greeks and the invading Persian army occurred in 490 BCE, and the second war was in 479–480 BCE.[5] In the first war, the Greeks were victorious, defeating Persian King Darius at the Battle of Marathon.[6] Part of the Greek victory in the Second Persian War was aided by the first well-recorded tsunami in the Eastern Mediterranean, which occurred in 479 BCE. It struck Potidea of Halkidiki in the Northern Aegean Sea, and it is reported that the waves of the tsunami destroyed the fleet of Persian King Xerxes.[7] This indicates that this was a seismically active period in the area. It also helped the Greeks win several decisive battles of Thermopylae, Artemisium, Salamis, Plataea, and Mycale, and the Spartans were by far the dominant force. This earned them the reputation as fierce warriors and helped cement the foundation of the Delian League treaty in 477 BCE among many of the Greek city-states of the time.

This neat political structure began to topple as the result of the Sparta earthquake. A very powerful earthquake with an estimated magnitude of 7.2 struck Sparta in 464 BCE. Based on the interpretation of description of the destruction from historical records, the ground shaking is estimated at X on the Modified Mercalli Scale of earthquake intensity which is described as extreme. This size earthquake is very unusual for this area, which typically has low-to-moderate seismicity. In fact, there has not been an earthquake even remotely as big since then.[8] The source of the earthquake was the 42-mile (68 km)-long Sparta Fault system just a few kilometers east of the city. A 12-mile (20 km) segment of the north–south-oriented fault slipped in normal motion, leaving a steep scarp or cliff on the surface some 33–40 feet (10–12 m) high.[9]

Because the area around Sparta was not used to such large earthquakes, the people and the infrastructure were completely unprepared for such an event. As a result, the damage was devastating. There are written accounts that the entire city was razed to the ground with not a single house escaping destruction.[10] Reports further describe that anything left standing was knocked over through a long period of aftershocks after the main earthquake including stone walls.[11] This is not unexpected because an earthquake of this magnitude most certainly was followed by strong aftershocks for an extended period. It is reported that more than 20,000 Spartan citizens perished in the disaster, although that number has been questioned recently. It is reported that many young, prospective soldiers in training were killed when the roof collapsed in the gymnasium where they were training. However, the dead were not only Spartan soldiers as many women and children perished in the disaster. This resulted in many fewer births among the Spartan soldier caste during the following years, which slowly but ultimately weakened Sparta's military power.[12]

But the earthquake disaster by itself was not the only event that changed history. It initiated a series of events that cascaded into a major change. At this time, the Spartans had enslaved a group of Messinian Helots to an agrarian existence designed to serve their Spartan masters. They were prohibited from possessing weapons and had very few rights and either lived in their master's homes on their estates or in small, controlled communities depending upon their role. When the earthquake struck, the Helots saw how much it had weakened the Spartans, so they took the opportunity to revolt. The weakened Spartans requested aid from Athens, possibly to help with damage from the earthquake, but more probably for help with quelling the rebellion. Athens sent troops who participated in laying siege to the Helots in their refuge at Mount Ithome. However, the Athenian presence in Sparta caused the Spartan aristocracy to fear that Athens might side with the Helots. As a result, after two years of assistance, the Spartans expelled the Athenians. There is a question as to whether this was done to show Spartan dominance over Athens, but either way, it offended the Athenians and drove them to explore ways to retaliate.

As a result of this insult, Athens broke its alliance with Sparta in 461 BCE and investigated other alliances as tension between the two intensified. After the rebellion in Sparta was put down, many Helots then fled to Athens who placed them around the strategically important Corinthian Gulf. Athens formed alliances with the city-states of Argos, Thessaly, and Megara to battle Corinthian expansion. They sent troops to Megara and built walls from the urban center to the port to prevent land invasion (Athenian long walls). Athens worked to establish the dominant naval force. These aggressive actions led to the First Peloponnesian War, which began in 458 BCE and included a number of battles and a reorganization of power. The Athenian–Spartan battles led the two to establish a treaty in 445 BCE.[13] The peace was short-lived when the second and much fiercer Second Peloponnesian War broke out in 431 BCE. This war lasted until 404 BCE with a crushing defeat of Athens by the Spartans, but it did not end the hostilities. An anti-Spartan Greek coalition developed in the region after the Peloponnesian War, and in the subsequent Corinthian War from 395 to 386 BCE, Sparta lost its naval dominance. Sparta finally lost its all military dominance by 370 BCE.

1755 Lisbon Earthquake

Perhaps it was the timing in history or the location in one of the great capital cities of Europe, but the confluence of several factors made the 1755 Great Lisbon earthquake one of the most impactful natural disasters in modern European history for Portugal, the rest of the continent, and even much of the world. The event even had ominous timing during the year, occurring on All Saints Day (November 1), at 9:40 a.m. when most Lisbon residents were attending mass in this overwhelmingly Catholic country. In fact, the Jesuits had a stronghold on the country at the time, and the inquisition was still operational. All of this would change as a direct result of the great earthquake.[14]

Even though there were other significant and even more powerful and damaging earthquakes in the region, this intraplate earthquake caused major historical and even scientific changes. It was generated from a thrust fault off the west coast of Portugal in the Atlantic Ocean[15] in a complex geological situation. The main structures passing through the western Mediterranean were generated by a large strike-slip fault zone called the Gibraltar-Azores transform fault. However, it was a horseshoe-shaped jog in this fault zone that likely ruptured producing this earthquake. The magnitude of the earthquake was first estimated at 9 which is among the strongest ever, but it was later reduced to 8.7–8.3 but more recently has been estimated at about 7.7.[16] Shaking is reported to have begun moderately at 9:35 a.m. and lasted for about 1.5 minutes. After a minute, a second violent shaking ensued and lasted for about 2.5 minutes. After another break of a minute, a final shock began violent but slowly softened over the following 3.5 minutes. Lisbon would never be the same after this short nine-minute event.

The shaking from this earthquake was felt at great distances from Lisbon. Several high chimneys toppled in Corunha, Spain, 500 miles (800 km) from the epicenter, and lamps that hung from church ceilings were reported swinging 750 miles (1,200 km) away.[17] The earthquake was felt as far away as Great Britain and Finland,[18] and seiches (sloshing waves) were reported in lakes, rivers, and harbors in Switzerland, England, Scotland, Finland, and Sweden. In Scotland, more than 1,200 miles (2,000 km) away, Lake Lamond experienced seiche waves of more than two feet (60 cm) for at least one and half hours. Similarly, the Dal river in Sweden more than 1,850 miles (3,000 km) away was also reported to have experienced seiche waves.

If the earthquake didn't do enough damage on its own, a massive tsunami was also generated by the earthquake, and it struck the coast of Portugal by about 11:00 a.m. further adding to the destruction. Many survivors of the earthquake rushed to the waterfront in Lisbon to escape the fires and collapsing buildings only to be killed and swept away by the massive wave, which rushed inland several hundred yards and flooded the first floor of remaining buildings. Like the earthquake, the size was initially overestimated at 400 feet (122 m) high. Actually, the tsunami was likely no more than about 66 feet (20 m) high at its peak on the Iberian coast, but it was very geographically extensive. The tsunami struck the Portuguese coast, southeastern Spain, North Africa, Great Britain, Cape Saint Vicente, the Algarve in the bay of Lagos, in Madeira and Azores Islands, and across the Atlantic Ocean in New Jersey,[19] making it a teletsunami. It even impacted coasts in the southern Atlantic such as Antigua, 3,600 miles (6,000 km) from Lisbon where it arrived at 7:30 p.m., Lisbon time. There, the highest wave was about 11.5 feet (3.5 m). It also struck Recife, Brazil, destroying many fisherman's huts. There are sixty-six records of tsunamis occurring along a 120 mile (200 km) stretch of the Spanish coast in the Gulf of Cádiz.[20] Although these waves caused extensive death and destruction and even proved that the east coast of North America was vulnerable to tsunamis, the tsunamis were not what changed history.

The grand capital city of Lisbon suffered immense destruction. The intensity of the shaking ranged from IX to VIII on the Modified Mercalli Scale in Portugal and VIII to VII in southwest Spain. The earthquake raised a huge dust cloud that shrouded much of the city and prevented people from seeing the ensuing chaos. The fires reportedly burned continuously for six days. For the first twenty-four hours after the main shock, there were so many aftershocks that it seemed like nearly continuous shaking. The first major aftershock occurred at about 11:00 a.m., and there were twenty-eight major aftershocks recorded during the first eight days. During the next six months, more than 250 aftershocks were felt, and there were 500 aftershocks recorded by September 1756.[21]

Between the damage from shaking and the massive fires it started, much of the city was in ruins. One source claims that only 3,000 of the 20,000 houses were inhabitable after the earthquake,[22] and others claim 85 percent destruction. About 38 percent of the larger churches were destroyed, and less than 10 percent had no damage. Almost 25 percent of the convents were destroyed with only about 5 percent undamaged. Hospitals were also devastated with about 20 percent destroyed, and 8 percent left undamaged. Of the public buildings, only small churches were less impacted with 65 percent undamaged. The Royal Palace was completely destroyed, but fortunately, the King and his advisors were away in Belem where earthquake damage was minimal. The newly opened Royal Opera House was destroyed as were the

Royal Mint, the Arsenal, and the Customs House. The harbor was destroyed by the tsunami leaving few boats operational. Lisbon residents were forced to live outside in tents for an extended period with minimal access to water and food as cleaning and rebuilding proceeded. The cost of all of the damage is estimated at 32–48 percent of the gross domestic product for the entire country of Portugal.[23]

The reported death toll for this disaster has also varied wildly through time. The population of Lisbon at the time is estimated at about 200,000 people. Of these, about 12,000 residents were killed directly by the earthquake. Including deaths outside of Lisbon and those killed by fires and the tsunami, the death toll is closer to 30,000–40,000. In addition, about 10,000 people are estimated to have been killed in Morocco and nearby areas of North Africa. A recent study on the local reports estimates the death toll in Portugal, Spain, and Morocco from the earthquake, fires, and the tsunamis at between 40,000 and 50,000 people, but the tsunami may have killed numerous others in coastal areas around the Mediterranean and Atlantic basins.

The changes to Portugal as a direct result of the earthquake were immense but primarily as the result of one man. For years prior to the earthquake, a political struggle had been gripping Portugal. Sebastião José de Carvalho e Melo, first Marquis of Pombal and later prime minister, was favored by the king but despised by the aristocracy. Prior to the Marquis of Pombal, all political and diplomatic offices were awarded to nobles, and the church had an iron grip on the country. The inquisition was still in control, and the Jesuits controlled the Catholic church. Pombal fought both groups to bring commoners into offices based on ability rather than birthright and to weaken the grip of the Jesuits on the population. The great Lisbon earthquake allowed Pombal to carry out his mission and to bring Portugal into a more modern stance than most of Europe.[24]

The King of Portugal had already given the Marquis of Pombal great power by the time the earthquake struck. But the earthquake so frightened the King that he retreated into seclusion, giving Pombal full authority in dealing with the emergency response and reconstruction. In fact, the King and his family stayed away from Lisbon during the emergency, living in tents outside of the city in the town of Belem for nine months. A few months after the earthquake, the prime minister died, and Pombal assumed his position as well. This gave him even more power, and he essentially ruled Portugal until the King died in 1777. Pombal first arranged the appointment of twelve district managers with emergency powers to quickly address rescue and recovery operations. He constructed gallows in several parts of the city to dissuade looting and executed thirty-four people for looting. He then worked with architects to design a well-organized city with a logical street plan to best accommodate commerce and transportation. All new buildings and rebuilding would be required to follow this plan. This slowed the rebuilding but ensured a more modern city.

This new power that the Marquis of Pombal gained as a result of the earthquake allowed him to be much more effective in his battle with Jesuits and the aristocracy. Pombal had his cousin hold a secret audience with the then Pope Benedict XIV to force reform and weakening of the Jesuits. He claimed that they were attempting to seize power from the King. After a failed assassination attempt on the King in 1758, Pombal wrote a widely disseminated treatise accusing the Jesuits of being behind the attempt and plotting to create their own separate state within Portugal. None of this is verified. Nonetheless, as a result, the Pope instructed the Cardinal-Patriarch to reform the church in Portugal. Pombal took the opportunity to arrest twelve Jesuits. In addition, troops were dispatched to surround all of the Jesuit colleges in Portugal. The situation quickly got out of hand, and on January 12, 1759, many Jesuits were accused of crimes against the crown, sentenced, and beheaded, broken on a rack, and burnt in Belem. These actions would later be condemned, and Pombal would have to answer for them. At that point, however, he took aim at the Inquisition. He stripped them of their policing powers and gave this authority to the public police department in 1768. The book censorship role of the Inquisition was also stripped and transferred to a royal commission. By the following year, Pombal also stripped the Inquisition's power as an independent tribunal. He ordered all of the property that had been confiscated by the Inquisition from Portuguese citizens to be transferred to the Portuguese national treasury. These actions effectively eliminated the Inquisition, which had terrorized the country for so long.

Pombal also removed many nobles from positions of power and replaced them with common people with talent. In this way, he broke the hold that the aristocracy had on the country. Unfortunately, in 1776, the King became gravely ill and subsequently died on February 24, 1777. Pombal lost his power, and all of the people who felt wronged by him took the opportunity to attack. He was accused of numerous wrongdoings and spent the rest of his life defending himself from his detractors. However, the changes he made were essentially permanent.

The Great Lisbon earthquake did more than just change Portugal; it changed all of Western Europe. Perhaps it was because of its location as the first major European city encountered by ships on the Atlantic routes or its role as a global trade center[25] but despite it not being the most powerful or deadly earthquake in southern Europe, its impact was immense. There were some dignitaries killed in the earthquake such as the Spanish ambassador, the Head of the English Seminary, and the great grandson of Racine, the French dramatist, but they could not account for the hysteria. The response was like an incident going viral on the internet, but invading all aspects of society and lasting for many years. It may have been the position in the age of enlightenment or the expansion of media, but the impact was universal.

Numerous accounts of the earthquake were published very quickly in almost every European language. They were published and republished in print from 1755 to 1757, and new accounts appeared regularly.[26] For example, on January 16, 1756, for no apparent reason, the Gazette de Cologne in Germany reminded readers that they were still thinking about the earthquake. Europeans started convincing themselves that they had felt the earthquake whether they had or not. Stories of survival in the disaster began appearing in major newspapers whether verified or not. Many of the accounts were sloppy and distorted with elements of fantasy and uncertainty. Some were purposefully exaggerated, falsified, and distorted to better appeal to readers. The more distorted the stories, the more popular they became, and they dominated conversation in public, private, and in the church and government for years.

The earthquake wasn't just harmless gossip either. This was a very tense time in Western Europe. It was still at the height of colonialism, but all available colonies had been divided up, and the riches they could provide were reaching their limit. The main players in harvesting wealth from colonies were Portugal, England, France, and Spain. England had attempted to ally itself with Portugal and maintained a presence in Lisbon much to the dismay of Spain and especially France. The French employed the publicist, spy, and adventurer, Ange Goudar, to investigate the situation in Portugal. He traveled to Portugal in July 1751 as part of a secret mission for the Secretary of Foreign Affairs and War but returned well before the earthquake. Nonetheless, he utilized the massive amount of circulating information about the disaster and combined it with his observations to quickly publish a book on it in 1756. The Seven Years War was primarily between France and England and primarily in the colonies but resulted in antagonism between them being high.[27] As a result, the book was an immediate success with four editions published in the first year alone, all in French. The account was full of errors, distortions, and propaganda and was more exaggerated with each revision. It wound up including an attack on the Portuguese Inquisition and anti-English propaganda. It claimed that Portugal was having its great Brazilian gold supply drained by the English through deceit and trickery. It raised the ire of the French who were jealous of the situation and it became a weapon of propaganda. This helped fuel the global Seven Years War, which had been brewing for years but which began in earnest in 1756. The war was between an alliance led by France against an alliance led by England but with very complex interrelations, suballiances, and subconflicts.

Portugal was too devastated by the earthquake to participate in the war and remained neutral as did Spain. This arrangement persisted for most of the war, but France grew impatient with Spain's neutrality and worked extensively to get them involved. In 1762, Goudar's book was published in Spanish and widely disseminated in Spain. The outrage in Spain as a result of this propaganda may have been the last straw to end neutrality. England declared war on Spain, and they, in turn, joined the French allies that same year. Later in the year, French and Spanish troops invaded northwest Portugal in an effort to encourage the Portuguese to join them and turn on the English. In response, England increased its aid to Portugal including troops, and the two soon drove out the invaders. The remaining allies from both sides from eastern and northern Europe either withdrew from the war or signed treaties to end their part in the war soon after. The fighting in the colonies also subsided, and the war ended in 1763 with the British making great territory gains in both North and Central America from those previously held by France and Spain.

The Great Lisbon earthquake occurred during the height of the Age of Enlightenment and was quickly addressed by some of the leading philosophers and intellectuals at the time. Having occurred on All Saint's Day added religious overtones to the entire incident and debate. First in a poem and later in his satirical work *Candide*, the French philosopher and writer Voltaire refuted the philosophy of "optimism," or the concept that God created a perfect world. He contemplated the nature of nature in contrast to religion and refuted the claim that Lisbon had suffered the disaster as punishment for a lack of piety.[28] In *Candide*, the Lisbon earthquake is used as the backdrop for Voltaire to criticize the optimism that was favored at the time by the intellectuals. Even though Rousseau was a friend of Voltaire, he strongly disagreed with the poem concerning the earthquake. He supported the Pope and further felt that humans created evil. He argued that if the residents of

Lisbon had not built such tall houses and so close together in such a dangerous area, the earthquake would not have been such a disaster. He further claimed that many residents were killed trying to protect their possessions rather than save their lives. It was this greed that made the earthquake such a disaster. The final great philosopher of the time who was involved with the debate about the earthquake was Immanuel Kant. Kant wrote two works about the earthquake.[29] In them, he agreed with Rousseau regarding the support of the Pope and that Lisbon should not have been located in an earthquake-prone area or built without safeguards. However, he agreed with Voltaire that this was a natural event and in no way punishment from God for sins committed. He studied the scientific data of the event on what was damaged and what wasn't to reach these conclusions.

The debate and fame of the earthquake did not end with the input of these philosophers but instead kept raging for several decades. More than a century later, the American physician, scientist, and poet, Oliver Wendell Holmes, wrote and published the poem "One Hoss Shay" in the Atlantic Monthly magazine in September 1858. This poem is an account of the Great Lisbon earthquake as experienced by regular residents and still captivated the interest of readers.

Yet another impact of the Great Lisbon earthquake was the initiation of methods for modern scientific seismology. Soon after the rescue and recovery operations ceased, the Marquis of Pombal conducted a survey of all the parishes of the country. The survey contained thirteen questions that were answered by as many people as possible. The questions included the time and duration of the earthquake, what direction buildings fell, the number of destroyed houses, how many people were killed, changes in surface water, ground fissures, the number and intensity of aftershocks, and the number of fires. The survey also asked about local populations, lack of food, and what measures were taken in rescue and relief.[30] The results of this survey were compiled to determine the distribution of damage and earthquake intensity. Eventually, this type of survey and analysis would evolve into intensity scales like the Modified Mercalli Scale, which is still used in earthquake analysis today.

1812 Caracas Earthquake

It may seem that earthquakes should be more powerful or kill more people to change history, but even moderately damaging earthquakes can be extremely impactful if they strike under the right conditions. This was certainly the case with the 1812 Caracas, Venezuela earthquake. After the American Revolution broke the stranglehold of colonialism in 1776, all colonies of the major European countries began moving toward their own independence. This was especially true in Central and South America, where the people had been under the rule of Spain and Portugal for 300 years. In 1800, all territories were still colonies, but this would soon change, and Venezuela was the first country to experiment with independence.[31]

On July 5, 1811, Generalissimo Miranda and the famous patriot Simon Bolivar led their troops into the capital city of Caracas and proclaimed independence. Over the next several months, the citizens built a new republic including a new federal constitution by December. General Miranda was assigned as commander of the army. The capital city of Caracas was beautiful, prosperous, and tranquil with no alarming revolts, no calamities, and not even much sickness. All Venezuelan citizens were guaranteed personal liberty and permitted property ownership. The previously common torture was abolished, and the centuries-old slave trade was condemned. The titles of nobility were abolished, and the Holy inquisition was eliminated. No property, crops, or items could be seized without provocation. Considering the laws and practices for the past three centuries, this whole new republic was starting out with security, prosperity, peace, and optimism.[32] Unfortunately, this idyllic scene was about to abruptly end.

Besides location and historical situation, timing is everything for a natural disaster to be impactful, and the 1812 Caracas earthquake was the perfect timing for it to change history. On Holy Thursday (or Maundy Thursday), March 26, 1812, at 4:07 p.m., just as most adult residents were attending church services, a strong earthquake struck the Caracas area. Based on the damage, the magnitude is estimated between 6.9 and 7.2, which is especially strong for the area. The source of the earthquake was the San Sebastian system fault, which is along the complex transform boundary that separates the Caribbean plate from the South American plate. This boundary accommodates the Caribbean plate moving east relative to South America. This transform is broad with many interconnecting faults along the Boconó–San Sebastian–El Pilar fault system, making it especially dangerous. Shaking from the earthquake caused massive destruction and as a result, intensity is estimated at up to IX–X on the Modified Mercalli Scale with broad areas ranging from VIII to IX.[33]

The earthquake was devastating in Caracas where the bulk of the population was located. However, cities of La Guaira (12 miles (20 km) north of Caracas), San Felipe, Barquisimeto, Merida, and several other major towns were also destroyed. Caracas had established itself as the economic, political, and religious center of the region over the previous several decades, receiving the appointment of the General Captainship in 1777, Royal Audience in 1786, Royal Consulate in 1793, and the Archdiocese in 1804, and it was named the capital of the Province of Venezuela. As such, the bulk of the population resided in the area, which means that the death toll was high. The problem is that many files were destroyed where record buildings collapsed, and subsequent lootings and official papers were used for fuel for bonfires or in ammunition. Therefore, the estimates on the death toll vary. The largest estimate is 20,000 people, but 13,000, 12,000, and 6,000 have also been proposed, and recent calculations indicate that about 2,000 people died in Caracas alone of the 32,000 residents.

The earthquake destroyed many of the buildings in Caracas. Estimates of the damaged and destroyed buildings range from 33 to 90 percent. One report claims that of the approximately 5,000 houses that stood before the earthquake, only about 2,000 remained standing after it. At the time of the earthquake, Caracas had four basic types of structures: churches, government buildings, homes of the wealthy, and homes of the rest of the residents. Of these, the churches were not as badly damaged, but the homes of the wealthy were destroyed because they had elevated, fragile features, and brittle building materials. Because the wealthy people were at church during the earthquake and their children were at home, there was a higher percentage of children killed in the disaster, as well as the servants who were attending to them.

This disaster was bad enough, but several confounding issues made it much worse. The young independent republic was not embraced by all Venezuelans. Many remained loyal to King Ferdinand VII of Spain, especially the clergy. They argued that the king was ordained by God and opposing him; they were going against God. However, even much of the general public who supported the republic were not sure that it would be able to survive. This lack of confidence coupled with outright opposition left the country with a weak resolve at best. The earthquake striking on Holy Thursday convinced the superstitious that they had made a mistake supporting the patriots. The clergy quickly took advantage of the coincidence and chastised the public even during rescue operations. They preached that now the people would feel the wrath of God for their disobedience. Simon Bolivar was also in Caracas and fully engaged in rescuing as many people as he could. The legend of this scene is that Bolivar came across a priest preaching repentance to injured people and threatened him at sword point to stop preaching and help save survivors. Bolivar's legendary status was partly earned during this disaster. He is quoted as having said "If Nature opposes us, we'll fight against her and make her obey us."[34]

The weak or lack of support for the republic by many of the residents was bad enough, but even worse, there were people taking action against it. Not long before the earthquake, Domingo Monteverde took it upon himself to return Venezuela to Spanish rule under King Ferdinand VII. Monteverde was a field marshal in the Spanish army who was seeking fame and position and decided that overthrowing the patriot republic would accomplish both. He assembled a small force, landed in Venezuela, invaded the city of Carora about 300 miles (500 km) west of Caracas, and defeated the patriots in early March of 1812. A force was sent by the republic to stop his advance at the city of Barquisimeto, and they succeeded. Monteverde's force had run out of ammunition, food, and supplies and was about to retreat. But then a miracle happened. The earthquake struck devastating Barquisimeto and killing two-thirds of the soldiers in the city barracks. As a result, Monteverde and his force entered the city on April 2 without firing a shot. They gathered all the arms, ammunition, and supplies they needed from the barracks to accelerate the offensive indefinitely. Even worse, Monteverde's troops now believed that God was with them in their cause in contrast to the republican troops who now doubted their cause. Further, many royalist, superstitious, and even unsure Venezuelans joined Monteverde further improving the attitude of his army.[35]

Caracas was in such disarray that they could not properly respond to this threat in a timely manner. The city suffered from critical shortages of food, supplies, weapons, and soldiers. The San Carlos Barracks in the city had collapsed during the earthquake, killing 600 patriot soldiers. The roads were so covered in rubble that supply vehicles, and rescuers could not pass through. In response, the city residents were ordered to evacuate a few days after the earthquake, and municipal workers began to clear the rubble. The church was left to conduct the remaining rescue and relief efforts. The residents set up tent cities outside of Caracas, and other destroyed cities set up similar arrangements. There were so many bodies that municipal workers initially dumped them into ocean, but the process went so slow that later they just burned the bodies forty at a time. Venezuela pleaded with European countries to send aid, but none would comply because they were dealing with the Napoleonic Wars and because the country was too unstable. The United States was sympathetic to their cause and sent $50,000 plus six ships loaded with food and supplies, but they wound up stuck in the harbor with no labor available to unload them. The republic requested to be recognized by United States as a sovereign nation, but the instability made them uneasy, and they refused. As a result, between the earthquake and the war, the country was broke and could not feed its people. Widespread famine ensued during April and May.

Monteverde took the opportunity left by the chaos to mount his attack. The leader of the revolution, Generalissimo Miranda, was named Dictator by Congress to put down this threat, and he took his army of 16,000 patriots to fight Monteverde. But Monteverde took city after city. He was stopped temporarily at San Carlos but soon triumphed and sacked the city. On April 4, a strong aftershock struck reinforcing the belief of divine intervention. It further strengthened the resolve of the royalists and weakened the patriots. On May 1, Miranda marched 12,000 troops against Monteverde near Caracas. However, public sentiment had so turned in Monteverde's favor that Miranda's troops were apprehensive. Even Miranda began to doubt his cause. Early in his march, he heard a loud explosion and ordered the army to halt. The sound was from the erupting volcano San Vicente and was regarded as such a bad omen by the men that they began to desert him. A whole company of men deserted to join Monteverde. Miranda became distraught. Monteverde launched a surprise attack against him, and his troops repulsed it, but he did not take the opportunity to counterattack and wasted an advantage. At this point, Miranda's troops began to lose faith in their leader.[36] As a result, within six weeks, all of western Venezuela was under the control of Monteverde, and in the east, an attack on the patriot army killed 600 men.

Miranda then received an urgent message from Simon Bolivar that the fortress at Puerto Cabello was being threatened but had no defense. He requested troops. Miranda sent no troops to reinforce the fortress, which contained needed arms and ammunition. On June 30, the commander of the prison at Puerto Cabello set the prisoners free, and with patriot deserters, they took the fortress and raised the Spanish flag. They then turned the fort cannons on the harbor and the city and forced them to surrender. Bolivar attempted to defend the city with just forty men, but they deserted him. Monteverde immediately marched toward the city. Bolivar quickly secured a ship and sailed to La Guayra. This occurred on July 5, 1812, just one year after the independence of Venezuela.

On July 30, 1812, Miranda escaped to La Guayra, condemned by the patriots for his treasonous actions. He was apprehended by patriots and anonymously turned over to the Spanish authorities for arrest. They sent him to Porto Rico and then to Cadiz, where he was locked in the prison at the fort of La Caraca for the rest of his life. He died in chains and miserable on July 14, 1816. Bolívar was sent into exile in Cartagena, but he was not away for long. In 1813, he was appointed commander of an expeditionary force to Venezuela to see if it could be retaken. Instead, he launched a sweeping campaign and defeated the Spanish in six intense battles. He retook Caracas and founded the Second Republic of Venezuela just a year later. However, he was soon driven from power yet again but returned repeatedly until he finally permanently freed Venezuela from Spanish rule in 1821.[37]

With momentum and a great reputation as a freedom fighter, Bolívar sent his top lieutenant to Ecuador to lead a great battle that gave them their independence just a year later in 1822. Just two years later, Bolivar and his lieutenant led two decisive battles that liberated a large area that became Peru and Bolivia, making him directly responsible for the liberation of four South American countries. Argentina already declared independence in 1816, Chile in 1818, and finally Brazil in 1825, making South America almost entirely politically independent. Although these other countries gained independence without Bolivar, they recognized the success he had in Venezuela as the incentive for their own victories. Bolívar clearly has historic importance as a liberator. However, it is widely recognized that the 1812 earthquake pushed him into this role. Bolívar himself recognized that without the earthquake, devastation of republican cities, and the resulting successful royalist backlash, the First Republic of Venezuela could have survived, and he would have remained there. The liberation of South America from Spanish colonial rule would have taken much longer. The lessons Bolivar learned from dealing with the earthquake formed his basic strategy of government during the subsequent long wars of independence.[38]

Meanwhile, because of the bankrupt treasury from the earthquake's destruction and the many wars, Caracas remained in a disaster condition for many decades. The government made several plans to rebuild it, but they were never carried out. Even after the patriots liberated the city in 1821 for the last time, nothing was done. It is reported that it took until the 1870s–1880s for the devastation from the earthquake to finally be repaired and rebuilt, up to seventy-five years later.

1923: The Great Kantō Earthquake

When it comes to being in the wrong place for natural disasters, Japan is among the most dangerous. Located on the Pacific Ocean "ring of fire," Japan is uniquely at the intersection of two interacting subduction zone plate boundaries. One subduction zone can produce devastating volcanic eruptions and earthquakes on its own, but two boundaries that interact add complexities that are difficult to imagine. As a result, Japan has been besieged by major earthquakes throughout its history, and because it is an island, it also has regular tsunamis as a result. Many of these earthquakes have been disasters, but none have changed Japanese and world history as profoundly as the 1923 Great Kanto earthquake, which destroyed the capital city of Tokyo and the surrounding area.[39]

The plate tectonic geology of the Tokyo Bay region of Japan is very complex. The epicenter of the Great Kanto earthquake was located slightly north of the

Sagami trough, where the Philippine Sea plate moves northwestward and subducts at an oblique angle beneath the island of Honshu. To the north, the Pacific plate drives westward, also subducting beneath Honshu. These two subducting plates converge below Kanto, where the Philippine Sea plate pushes above the Pacific plate. The earthquake focus was within the Sagami trough on faults showing a combination of reverse and dextral strike-slip of 19 feet (6 m) on one plane and 31 feet (9.5 m) on another.[40] These faults are parts of a megathrust of the Philippine Sea plate, and the area of rupture is estimated at 60 miles (100 km) by 60 miles (100 km).

The Great Kanto earthquake struck the Sagami Bay at 11:59 a.m. on September 1, 1923. The epicenter was approximately 60 miles (100 km) from central Tokyo and 30 miles (66 km) southeast of Yokohama. The depth of the focus was an estimated 14 miles (23 km), which is relatively shallow for a subduction zone earthquake, making it more damaging than most. The main shock struck was so powerful that it badly damaged the Tokyo seismograph, releasing energy equivalent to 400 Hiroshima-sized atomic bombs. As such, the magnitude can only be estimated between 7.9 and 8.2. The duration of the main shock can also only be estimated at four minutes, but the range from eyewitness accounts ranges from forty-eight seconds to ten minutes. The surface wave intensity is estimated at XI on the Modified Mercalli Scale, which is extreme.[41] Immediately after the main shock, two large aftershocks of estimated magnitudes of 7.2 and 7.3, respectively, struck the southern Kanto district. The first aftershock occurred 3 minutes after the main shock, and the second aftershock occurred 4.5 minutes after the main shock.[42]

The earthquake was accompanied by rapid uplift and subsidence in the Sagami Bay. The Miura Peninsula and Boso Peninsula on the north side of the Sagami trough had uplifts of 10–30 feet (3–9 m).

On the east coast, of the Izu Peninsula to the south of the Sagami trough, the ground surface was depressed 1.5–5.9 feet (0.45–1.8 m) in one area but uplifted 6.4–9.4 feet (1.95–2.85 m) in an adjacent area. These bathymetric and topographic changes were permanent. The vertical surface motions generated tsunamis across the bay up to 20 feet (6 m) high,[43] although some sources report 40-foot (12 m) tsunamis.[44] They killed at least 100 people and washed away more than 100 houses.

The surface waves caused severe shaking of the ground surface. The volcanic soils and steep slopes could not withstand this shaking. As a result, there were an amazing 89,080 slope failures over an area of 33.3 square miles (86.3 km^2) in the Tokyo-Yokohama area. This constitutes at least 7 percent of the forested area in the region.[45] These slope failures destroyed houses and other structures in their path.

The main shock and aftershocks, slope failures, surface shaking, and tsunamis did enough damage on their own to make the Great Kanto earthquake a major disaster. However, the main shock and immediate aftershock struck at noon. In Tokyo, many people were cooking lunch primarily on open fires in small braziers, which were overturned by the shaking, igniting some 130 separate fires, and the flames spread to the largely wooden houses.[46] Fires quickly engulfed both Tokyo and Yokohama. Unfortunately, a typhoon was passing off the coast of the Noto Peninsula in Northern Japan at the time, which produced little rain but winds nearing 60 miles per hour (97 kph), driving the flames and creating horrifying firestorms. The shaking severed the city water lines, and fire brigades could not get through the debris-covered streets even if there was water to extinguish the fires. These multiple contributing factors allowed the fires to develop into an inferno. In Tokyo, the fire was so hot that it formed firestorms including a 300-foot (95 m) high fire tornado that was so powerful, and it lifted people and large objects high into the air and threw them. Panic-stricken survivors rushed to an army depot and an open area in a city park, which was away isolated from the burning buildings. However, a firestorm swept quickly into the area and killed 35,000–40,000 people, leaving only a few hundred badly burned survivors.[47]

The total death toll for this horrifying disaster is estimated between 105,000 and 142,800 people. It left both Tokyo and Yokohama in ruins. The strong shaking damaged at least 75 percent of the buildings in Tokyo, but the fire completely incinerated 67 percent of the city as it burned for three days. By the morning of September 3, 6 square miles (18 km^2) of Tokyo were incinerated. This destruction left as many as 2.5 million people homeless.[48] Yokohama was in equally horrific condition with nearly every building destroyed. The famous Yokohama Grand Hotel, which had hosted famous people like US president Taft and author Rudyard Kipling, completely collapsed, killing hundreds of hotel employees and guests. Within a month of the earthquake, Tokyo was a city of barracks, temporary shelters, and makeshift communities within the ruins of the destroyed city.[49]

As is usually the case, it is not just the location of a disaster to change history; it is also the timing. Tokyo is the capital of the powerful country of Japan, but the timing was key at many scales. The ruling prime minister died one week before the earthquake, leaving the government essentially leaderless during the disaster. The eventual successor was still choosing a cabinet at the time. As a result, the Minister of Foreign Affairs who had been appointed provisional prime minister issued an Emergency Conscript Law the next day. The emperor declared martial law that day, and the new prime minister and his cabinet were rushed into place by late in the day of September 2.[50]

However, trouble had already begun. Originating in Yokohama, rumors began that bands of lawless Koreans, who were an ethnic minority in Tokyo, were setting fires and looting shelters, stores, and vacant homes, and even poisoning water wells throughout the Kanto area. By the morning of September 2, these rumors of Korean lawlessness had ballooned into bombings that caused the fires and that thousands of Koreans were heading toward Tokyo to wreak havoc there. The rumors caused panic and hysteria in the Japanese civilian population. They organized into groups of as many as 1,600 vigilantes to patrol the ruins under the pretense of maintaining public safety. However, a number of the vigilantes turned violent, hunting and killing innocent Koreans. There is disagreement on the number of Koreans killed by the vigilantes with the police downplaying it at 231 deaths but independent sources finding that 6,000 or more Koreans were killed. The government condemned these actions, stating not only that the attacks undermined their goal of assimilating the Koreans into Japanese culture but also that they brought shame to Japan. On September 5, the newly installed prime minister of Japan issued a statement condemning the actions of vigilantes toward Koreans.[51]

The new government quickly established an Emergency Relief Office and charged it with the task of providing food, shelter, and medical assistance to the survivors. In order to maintain martial law, they also called in the military. By September 8, they had mobilized 35,000 army personnel to the Tokyo area including twenty-one infantry regiments, six cavalry regiments, eighteen engineer battalions, two railway regiments, two telegraph regiments, flying corps, and medical corps. They played the primary role in maintaining order and providing relief to the survivors and were viewed with reverence. On September 12, the Japanese Prince Regent issued an imperial edict defending these actions because it was an extraordinary time that demanded extraordinary actions.

The disaster also played an integral role in the long-term development of Japan and the world. The Allied victory in the First World War encouraged Western values to be integrated into Japanese culture. Japanese citizens wanted more popular representation in government. Advocates of a liberal democracy in Japan were encouraged in this new order, and by 1919, the working class had joined with intellectuals and students a growing movement. In 1918, the first prime minister to head a majority party cabinet supported and encouraged these values. Society in Japan became much more materialistic and individualistic with the working class that was gaining political power. The country even had to deal with a labor strike among rice workers in 1918, which would have been unheard of prior to this time. The working class occupied the lower flatlands of the city in contrast to the established wealthy who occupied the higher hilly areas. This shift in power caused friction between the two groups.[52] The liberal movement was already in trouble when the supportive prime minister left in 1921. But the Great Kanto Earthquake put an end to this political shift by imposing much more damage to the low city than the high city.

The established, conservative Japanese leaders took advantage of the disaster to swing the political power back to them. Many of these Japanese elite, from politicians to religious leaders, journalists, and social commentators, declared the earthquake disaster to be divine punishment for the sins of society. The decadence, extravagance, selfishness, frivolity, individualism, and the pursuit of luxury that dominated the 1920s in urban Japan were not in the Japanese values and needed to be abandoned.[53] These people attempted to change the public view of the earthquake from an agent of destruction to one of the salvation. They argued that the social and political disaster they had been facing prior to the earthquake was much more dangerous. This campaign was so successful that a public survey conducted in 1924 found that most people regarded the earthquake as a divine punishment, a divine warning, or a divine opportunity to reform society.

The new prime minister and his cabinet devised long-term plans that utilized the earthquake as an opportunity to change social and political trends. As early as September 4, the newly appointed Japanese Home Secretary drafted a plan to reconstruct the city of Tokyo and submitted it for review by September 6. This plan included a complete transformation of Tokyo into a city of the future, utilizing state-of-the-art planning, architecture, and building materials. This new imperial capital would emphasize state power and authority. This transformation was also built around the suppression of the liberal view and convincing the public to consent to conservative rule.[54] To support and expand this effort, on November 11, the Emperor issued an edict encouraging a "Sincere and Strenuous Life" among the Japanese people. The government had decided not only to reinvigorate Tokyo but also the entire nation.

Their first and major effort of this plan was to overhaul the education system. Within three months of the earthquake, the Ministry of Education published a collection of stories describing ordinary people helping others and saving lives. These stories were designed to glorify service to the people and the nation as the highest form of service and to change society. The objective of infusing these stories into the education curriculum was to serve the government's effort to instill their moral values into school-age children through capitalizing on the disaster. To support this effort, state agencies launched moral campaigns to the public to convince the Japanese people to be frugal, loyal to the Emperor and country, and to be earnest and obedient. The politicians also implemented a series of outrageously steep antiluxury taxes on goods and services that were deemed extravagant or wasteful.[55]

Through this onslaught of coercion based around the Great Kanto earthquake, the government quickly transformed the priorities of the country to loyalty to the Emperor, filial piety, personal sacrifice, courage, and obedience. The right-wing architects of the plan were quick to claim victory when Japan weathered the 1929 worldwide economic crisis relatively well. However, they also engineered the Emergency Japan of the wartime 1930s. They even staged the fake Manchurian Incident of 1931 to allow them to invade Manchuria. This exercise prepared the Japanese people to mobilize human and natural resources for war.[56] In essence, the Great Kanto earthquake was used as a pretense to build Japan's military might and public opinion ultimately to their nationalism and their aggressiveness in the Second World War.

5

Hurricanes

Hurricanes, cyclones, and typhoons are powerful storms that are among the deadliest natural disasters. These storms originate in the tropics and develop into the most powerful meteorological events on earth. All are produced through the same processes and are equally powerful but classified by location. They are hurricanes in the Atlantic Ocean and adjacent waters, but they are typhoons in the Pacific Ocean and cyclones in the Indian Ocean. A fully developed hurricane can produce the energy needs of the United States for an entire year in a single day.[1]

Hurricanes begin as weakly developed innocuous clusters of thunderstorms off the west coast of Africa in the Atlantic Ocean. They develop in large stable air masses that are normally associated with pleasant weather. The surface ocean water temperature must be 79 °F (26 °C) or higher for strengthening. This temperature elevates the humidity level in the air above the ocean surface to a critical point. The warm, moist air is drawn into updrafts in the thunderstorms. As the air is swept upward, the moisture condenses, and the condensation process releases latent heat energy into the storm. This energy powers up the storm, causing more vigorous circulation and resulting in stronger updrafts. The upward-moving air reduces the air pressure in the storm, making it an ever-stronger low-pressure system. The intensifying low draws increasingly stronger winds across the ocean surface and into the storm. These winds whip up the surface of the ocean into ever larger waves that expose more surface area on the ocean surface from which more humidity can be emitted. The humidity, in turn, results in more condensation and more latent heat transfer. This positive feedback process quickly powers up the storm.

These storms rotate counterclockwise in the Northern Hemisphere but clockwise in the Southern Hemisphere, and they cannot cross the equator. As hurricanes grow, they form a clear and calm eye at their center that typically ranges from 20 to 40 miles (32–64 km) across. The eyewall which surrounds this eye is the most dangerous part of the storm, containing the strongest winds and heaviest rains. Rainbands of 5–10 miles (8–16 km) width and several hundred miles in length surround the eyewall storm in a spiral pattern. Hurricanes, cyclones, and typhoons are usually about 300 miles (480 km) across but can range from 200 to more than 500 miles (320–800 km). Strong winds persist between the rainbands, but there is little to no rain. The heavy clouds and rain occur where air is rising in a hurricane, but there is no rain where the air is falling.

Hurricanes lack atmospheric guidance, in many cases, because they are not connected to fronts. This makes their directions and travel speeds difficult to predict and is the reason why hurricanes can follow unusual paths. In general, hurricanes travel at about 15–20 miles per hour (24–32 kph), but they can sit stationary for extended periods or achieve forward speeds of 60 miles per hour (96 kph) in some cases. Hurricanes are easily degraded if they make landfall, cross the ocean with surface temperatures below 79 °F (26 °C), encounter high-level winds, or encounter a front.

The main classification system for tropical disturbances divides them into tropical depressions, tropical storms, or hurricanes. Tropical depressions are systems with organized thunderstorms, surface circulations, and sustained winds less than 38 miles per hour (61 kph). Tropical storms are similar but with strong thunderstorms and sustained winds of 39–73 miles per hour (62–117 kph). Hurricanes have the strongest thunderstorms, well-defined circulations, and maximum sustained winds of 74 miles per hour (118 kph) or more.

The Saffir–Simpson Scale is used to classify hurricanes in five categories based solely on wind speed. Category 1 hurricanes are the weakest with sustained winds of 74–95 miles per hour (119–153 kph) and the least damaging. Category 2 hurricanes have sustained winds of 96–110 miles per hour (154–177 kph) and include coastal flooding and damage to roofing. Category 3 hurricanes have sustained winds of 111–130 miles per hour (178–209 kph) and produce structural damage to small buildings and coastal flooding up to 8 miles (13 km) inland. Category 4 hurricanes have sustained winds of 131–155 miles per hour (210–249 kph) and produce removal of roofing of most buildings and require evacuations of areas with elevations below 10 feet (3 m) above sea level. Category 5 hurricanes have sustained winds greater than 155 miles per hour (249 kph) and can produce complete building failures and major flooding damage to lower floors of structures less than 15 feet (4.6 m) above sea level.[2]

The Hurricane that Saved the United States

Americans celebrate Independence Day each year in celebration of the declaration of independence on July 4, 1776. However, securing that independence for the entire fledgling country was not so easy. Although there were some major victories, the British Navy was unstoppable on the seas, and they quickly secured the most important coastal areas and held them throughout the war. By 1780, the British held New York, controlled the Chesapeake Bay region, and handed the Continental army their worst defeat in Charleston, South Carolina where Lord Cornwallis was amassing forces to march north and crush the remainder of the Continental Army. That force was reduced to 3,500 hungry, under-supplied, and unpaid troops who were always on the verge of mutiny and desertion.[3] A grim situation indeed.

In addition to the situation in the Revolutionary War, conflict extended into the Caribbean, where the hurricanes developed and struck. The Seven Years War ended just a few years earlier in 1763, and although it involved much of Europe, the tensions among the major colonial rulers of the area including France, Spain, Great Britian, and Portugal persisted. Most of the Caribbean islands and even the land around them were divided among Spain, France, and Great Britain. For example, Cuba was a Spanish colony, whereas Haiti belonged to France, and Jamaica belonged to Great Britain. France and Spain had a loose alliance against the British in the region, which erupted into regular naval battles and even some land conquests. The conflict between the French and British was so onerous that France was supplying and offering assistance to the Americans in the Revolutionary War. Otherwise, the Caribbean region was a goldmine to the colonialists, producing sugar, rum, and many other agricultural products largely through plantations heavily stocked with African slaves.[4] It was the greed of the European nations that drove the conflict.

Hurricanes can cause terrible natural disasters, so even though the United States was helped, many people suffered and died at the hands of the events of 1780. The entire hurricane season was bad in 1780, but late-season conditions developed to perfectly spawn several killer storms, making October of 1780 the deadliest month for hurricanes in recorded history. Three deadly storms developed and killed at least 27,500 people across the Caribbean[5] and wreaked havoc among both the Europeans and slaves alike, neither of whom had experience with such massive storms. This single month was a sobering experience for these relative newcomers, and it changed the history there as well as the United States.

The first storm has been named the Savanna-la-Mar hurricane. With no warning, it made landfall in Jamaica on October 3. accompanied by a swift 20-foot-high (6 m) storm surge that swept away a large group of people who had gathered to observe the ominous clouds. The storm crossed Jamaica, destroying the port of Savanna-la-Mar and the towns of Lucea and Montego Bay while flattening sugarcane fields and destroying homes and buildings. The storm headed northwest after exiting Jamaica, crossing directly over Cuba and the Bahamas, and wreaking havoc there as well. Over the ocean and ports, however, it destroyed numerous ships. It destroyed seven British Navy ships including the 64-gun Stirling Castle and the 44-gun Phoenix killing most of their crews. The Phoenix was struck by the storm between Jamaica and Cuba. The storm was so intense that it drove the Phoenix into the shore just east of Cape Cruz, Cuba, so high onto the rocky shoreline that it could not move. The ship was a total loss, although only five sailors had been killed. Other ships were not as lucky. For example, a British prisoner transport ship was sunk, killing several hundred Spanish prisoners. In all, approximately 3,000 people were killed in this hurricane, making it among the more deadly storms.[6]

However, the Savanna-la-Mar hurricane was minor compared to the monster that was about to strike. The Great Hurricane of 1780 or San Calixto Hurricane devastated the Antilles and West Indies from October 9 to October 16, 1780. It is speculated that the storm began in the Cape Verde Islands off the west coast of Africa and strengthened while drifting across the Atlantic Ocean and reaching hurricane strength southeast of the Lesser Antilles. It made landfall on the island of Barbados late in the day on October 9 at nearly peak intensity. Damage reports submitted by naval officers and experimental simulations indicate that the hurricane had sustained wind speeds of over 200 miles per hour (320 kph), making it a strong category 5 storm. Reports indicate that the hurricane sat over the Barbados for nearly two grueling days, with the eye just north of the island. The storm completely demolished St. Michael's, St. George's, Christ Church's, and St. Lucy's churches, as well as St. Thomas's Chapel. Trees were uprooted and stripped of their bark all across the island, and every house and fort was damaged or destroyed. It is further reported that strong winds lifted heavy artillery at least 100 feet into the air.

All of this destruction was accompanied by the deaths of at least 4,500 people on Barbados alone.[7,8]

The storm finally began moving very slowly northwestward near St. Lucia and Martinique, still maintaining its category 5 strength before passing over the island of Mona between Puerto Rico and Hispaniola. The storm then swung northeastward near 70° W longitude on October 14 before moving southeast of Bermuda and into the open Atlantic Ocean on October 16 but leaving a path of total destruction along the way. The northwest coast Martinique was struck by a 25-foot (7.6 m) storm surge that decimated the town of Saint-Pierre, causing more than 9,000 deaths and destroying all of the buildings. On Saint Lucia, a British ship was lifted from the water and dropped onto the hospital at Port Castries. The hurricane is reported to have destroyed 97 percent of the buildings in the city of Kingstown on the island of Saint Vincent. Heavy rain and winds hit Antigua and Saint Kitts, driving many ships onshore. At Grenada, nineteen Dutch ships were wrecked. About fifty ships were wrecked in Bermuda. Fortunately, Puerto Rico and Hispaniola were on the margins of the hurricane but still had heavy rainfall, high winds, and flooding. The worst damage was in Barbados, St. Lucia, St. Vincent, Martinique, southern Guadeloupe, St. Kitts, St. Eustatius, and lesser amounts in southwestern Puerto Rico, and southeastern St. Domingo.[9]

The ferocity of the storm caught everyone by surprise and crippled the sizable naval forces in the area. The British Navy suffered the worst of the damage with fifteen warships either sunk or severely damaged. In Martinique, the Andromeda, Edymion, and Laurel were destroyed, and most of their crews were drowned. In St. Lucia, five British warships were destroyed. Other British warships lost, including all crews, were the Blanche, the Thunderer, the Stirling Castle, the Scarborough, the Barbados, the Deal Castle, the Victor, and the Endeavor. In addition, seven British naval ships were badly damaged including the Egmont. The French Navy had a force of forty ships involved in the American Revolutionary War effort off the island of Martinique that were in the path of the hurricane. However, only one ship, the frigate Junon, was lost. Several hundred soldiers were lost in the storm, but about 9,000 civilians were killed as well.[10]

The official death toll for the Great Hurricane is 22,000 people, which makes it the deadliest hurricane in history. However, some studies estimate that the actual death toll may be 27,000 or even 31,000 people. There is no estimate on the direct monetary cost of the disaster.

The British Navy was not the only one devastated by a hurricane but it was a different hurricane that came to their rescue. The Spanish saw that the Great Hurricane so damaged the British fleet in the Caribbean region that they would be unable to defend themselves. At the urging of Spanish General Don Bernardo de Galvéz, the Admiral Don José Solano y Bote Carrasco y Díaz decided to take advantage of the situation by attacking Pensacola, the capital city of British West Florida and the last outpost of the British on the northern Gulf Coast. He amassed 4,000 crack troops to sail on a sixty-four-ship armada with seven ships of the line, six frigates, two hospital ships, and over fifty transports loaded with supplies and led by Field Marshal Don Bernardo de Gálvez y Madrid, from Havana, Cuba. This massive invasion fleet left the Havana harbor, Cuba on October 16, anticipating no opposition.[11]

The Spanish were correct that they would encounter no military opposition, but they did not count on yet another deadly hurricane. After all, the Great Hurricane had yet to even exit the area. However, a third major hurricane in two weeks first brushed the coast of Jamaica on October 15. It appears to have traversed the Yucatan Channel on about October 18. It then overtook the Spanish armada somewhere west of the Dry Tortugas on October 20 and destroyed it thereby eliminating the attack.[12] Survivors report having fought howling wind and huge waves for 80 hellish hours. The crews jettisoned all of the cargo from provisions to arms to lighten the ships. About half of the troops and countless ships were lost in the storm. Solano and Gálvez limped their ships and most of those that remained back to Havana. Other crippled ships began arriving in ports across the region for the next several days. On

October 28, the crippled Jesús Nazareno schooner limped into Mobile Bay, Alabama. Four more crippled ships arrived over the next four weeks. It might seem that this storm called Solanos Hurricane changed history by saving Pensacola but the Spanish regrouped and successfully attacked and took the city the following spring. So, it really just delayed history.

However, the Great Hurricane did change history during the following year. Locally, agricultural production was abysmal. Thousands of slaves had been killed in the hurricanes, leaving a labor shortage. This was remedied with a brisk upturn in the slave trade over the following years. Sugar production had already declined because of the American Revolution, but it sharply declined in 1781 and did not return to pre-hurricane levels until 1783. Rum exports decreased 80 percent in 1781 and did not recover completely until 1784. Many landowners were forced to take loans, and some went bankrupt but far less than could have. Surprisingly, and in contrast to the way the Americans were treated, when landowners solicited assistance from Great Britain, they responded magnificently. The devastation of the 1780 hurricanes was met with an unprecedented campaign of humanitarian relief. This marked a big policy shift for Britain.[13]

The real impact on history, however, was in the American Revolution. Realizing that the devastation of the hurricanes put them at a great military advantage, the French repaired their ships and strengthened their hold on their Caribbean colonies during the winter of 1781. In the spring, they sent forces north to aid the struggling American colonies. The five years of war, inflation, and reduced trade had reduced the size of the Continental army and left many of the troops underpaid, disillusioned, and even mutinous in some cases. In contrast, the British had occupied New York for five years, and they had recently handed the Americans their worst defeats of the war in and around Charleston, South Carolina. To that point, the French Navy had made minimal input to the war effort since France joined the war three years earlier. This changed in the spring of 1781 when a French fleet of warships commanded by Comte de Grasse sailed from France to the Caribbean, picked up part of that fleet of ships, and sailed an armada to the Chesapeake Bay. The French fleet under Admiral de Barras in Newport joined the armada in Virginia, and General Rochambeau's troops joined with Washington's to attack the British in the south. When the British learned of the planned southern campaign, they sent a fleet of ships from New York to the Chesapeake to stop them. However, some ships were still damaged from the hurricanes and needed repairs, which slowed the trip. They finally engaged Comte de Grasse's fleet in the Chesapeake Bay on September 5. The French had twenty-six ships to the British nineteen ships, and after two hours of fighting, the French damaged enough of the British fleet to force them into retreat. They surrendered the Chesapeake Bay and sailed back to New York.[14]

The French victory at the Battle of the Chesapeake allowed the allies to contain British commander Lord Cornwallis at his Yorktown base and prevented the British Navy from resupplying his army. The combined French-American southern campaign advanced south and ultimately achieved victory over the British in the Battle of Yorktown in 1781. Even further exacerbating this situation, the British diverted naval forces from the Battle of Yorktown to Jamaica because the French had also decided to capitalize on British naval and military weakness there. Without any relief, Lord Cornwallis quickly surrendered to George Washington on October 19, 1781, which effectively ended the American Revolution. Subsequent peace negotiations resulted in the Preliminary articles of Peace, which was signed in Paris on November 30, 1782. However, none of this quick and resounding victory by the joint American and French armies would have been possible without the crippling of the British fleet in the Caribbean by the multiple massive hurricane disasters.

1900 Galveston Hurricane

The devastating Galveston hurricane of 1900 caused the worst natural disaster in history of the United States. The city of Galveston, Texas, sits on Galveston

Island beach facing onto the Gulf of Mexico. This beautiful location was part of the impetus to grow the city into a picturesque destination, especially after the Civil War. Sitting directly on the Gulf, the economy boomed through the shipping of cotton, sugar, molasses, cattle, pecans, and animal hides. The port shipped out 82,000 bales of cotton in 1854, but it increased steadily to 1878 when it ranked third in cotton shipments among all US ports. With all of this commerce, the population of Galveston boomed ahead of other Texas cities with 13,818 residents in the 1870 census and 22,248 in the 1880 census. By 1900, however, the advancement in transcontinental railroad transport and booming manufacturing in other parts of Texas allowed San Antonio with 53,321 residents, Houston with 44,633 residents, and Dallas with 42,638 residents to bypass Galveston with 37,789 residents.[15]

One limitation of Galveston as a port was the shallowness of the sandy harbor. To remedy this, the city attempted several channel deepening projects that failed. In 1874 and 1880, the US Army Corps of Engineers took over the effort, attempting to deepen the harbor to 20 feet (6 m). Finally, in 1890, the US Senate approved the Galveston Harbor Bill, deepening the harbor and installing jetties to keep out drifting sand. As a result, in October 1896, the largest cargo ship in the world was able to dock at Galveston wharves. This new ability immediately increased imports by 37 percent but exports also boomed. Galveston shipped out 22 percent of the cotton from Texas before the harbor-deepening project, but 64 percent in 1897–8 after it was deepened. In 1900, cotton shipments from the port at Galveston exceeded 2 million bales.[15] This made Galveston second only to New Orleans in US cotton shipments by 1900. In 1899–1900, more than $86 million in foreign exports passed through the booming Galveston port.[16]

All of the shipping money transformed Galveston into the grandest city in Texas by the turn of the twentieth century. It had the first telephones and electric lights, the most millionaires living in the grandest mansions, and the most exotic bordellos.[17] Tourists flocked to the grand city for the opulence, the broad beach, and sweeping ocean vistas. They enjoyed the new opera house and grand hotels, all built with cotton money. By 1900, the city had four national banks, thirty hotels, and thirty-five churches.[18] Galveston was called the New York of the Gulf Coast. It even became the destination for immigrants to the point that from 1890 to 1915, it was called the Ellis Island of the West.[19]

However, this grand city was not well thought out. It was located on a low, flat, barrier island of 30 miles (50 km) in length and less than 3 miles (5 km) in width that was really no more than a sandbar. The highest point was only 8.7 feet (2.6 m) above sea level, and most of the island was only a foot or two above sea level. Further, Galveston Island originally had natural protective barrier sand dunes in the seaward side of the island, but they had been removed so the sand could be used as fill in the low-lying parts of the city.[20] Considering that the area had been struck by hurricanes before and hurricanes were striking other parts of the Texas coast during this time of growth, locating such an important port with such an investment of wealth in such a precarious place was the recipe for disaster. That is exactly what happened.

It is assumed that the Galveston hurricane began its life like most other hurricanes as a weakly organized low-pressure system off the west coast of Africa sometime in mid-August of 1900. It slowly strengthened and drifted across the Atlantic Ocean before being detected as a tropical storm on August 27 by a passing ship to the east of the Windward Islands.[21] The storm moved west-northwest and, on August 30, was recognized by the US Weather Bureau as it entered the northeastern Caribbean Sea. It passed the Dominican Republic on September 2 and approached Cuba on September 3. When it soaked Cuba the next day, the Galveston weather station was alerted to its existence. The Cuban weather service predicted that the storm would move west but the US Weather Bureau chief in Washington D.C. disagreed with their assessment and did not issue an alert for the storm in the Gulf of Mexico. However, the storm entered the Gulf of Mexico on September 5 after crossing southern Florida and essentially disappeared from tracking. It apparently strengthened considerably as it sped across the warm Gulf waters, and local reports of its distant effects were ignored. In fact, the forecast for Galveston on Saturday, September 8, was for light rain and northerly winds.[22]

However, on September 7, the suspicious weather of growing winds, clouds, and rough seas led Galveston Weather Station Chief Isaac Cline to issue a hurricane alert, and he flew hurricane warning flags even though he did not have permission from the Washington D.C. office.[23] This warning was ominous because Cline stated that there was very little chance that a strong hurricane would hit Galveston in an article in the Galveston News newspaper in July 1891. But Galveston residents paid little attention to the deteriorating weather or the warning. No one evacuated, and onlookers even made a special visit to the beach just to see the large waves.

The sky began clear on the morning of September 8, but the large waves had flooded several blocks of downtown Galveston nearest to the water. Minor flooding was not uncommon, so residents also ignored this development and went about their regular business. However, by 8:00 a.m., the sky was overcast with frequent drizzle, and the wind had become stronger. By 10:00 a.m., the waves were much larger, and crowds had gathered to watch them. Cline contacted the US Weather Bureau in Washington, D.C., asking them to declare a hurricane warning, but at 11:15 a.m., he was told that the storm he was witnessing would stay out to sea.[24,25] He grew anxious about the lack of concern for this serious situation. He left the weather office and, with his brother, attempted to warn the residents closest to the beach of the impending danger and to seek shelter, but to no avail.

By early afternoon, the weather conditions had deteriorated to the point that all residents were taking the warnings seriously and seeking shelter. At around 2:30 p.m., the weather bureau rain gauge was blown off the roof. The bureau anemometer was swept away by the wind at about 6:00 p.m. after recording wind speeds of more than 100 miles per hour (161 kph).[26] The wind speed continued to increase, sending deadly projectiles of wood and slate from rooves all across the city and causing significant damage to buildings. The real catastrophic part of the hurricane, however, was the storm surge. The storm surge is water plowed by the wind into a bulge on the ocean surface. When the storm makes landfall, the sea level is raised by the height of the surge. In the Galveston hurricane, after the morning flooding, the flood waters rose continuously from about 3:00 p.m. to about 7:30 p.m., indicating incursion by the storm surge. It is estimated that the peak of the surge was 15.7 feet (4.8 m) high, which was significantly higher than the elevation of Galveston Island, meaning that it was entirely underwater. The hurricane intensity increased into the night, and the horrified residents faced the storm perils in pitch darkness. It is estimated that the storm was a category 4 hurricane by then with maximum sustained winds of 145 miles per hour (235 kph).

There are many harrowing stories of survival and death during the overnight. Cline allowed forty to fifty neighbors to shelter in his sturdy house where he thought they would be safe, but a bridge broke loose in the storm and rammed it, killing several of them including his wife. At least 700 people abandoned their homes to shelter in the town hall, but the roof collapsed and many died. Many people sheltered in a major downtown hotel but were forced to scramble up the stairs to escape the rising flood waters. Some survivors were crammed into the winding staircase of a tower to escape the flood. However, by far, the most heart-wrenching tragedy of the storm occurred at St. Mary's orphanage where ten Catholic nuns and ninety children died when waves destroyed two wooden dormitories. These buildings were built near the beach because the belief was that ocean breezes reduced the occurrence of yellow fever. The nuns tied the orphans together with a clothesline to keep them from being swept away, but all drowned and were found still tied together.[27]

The morning light brought apocalyptic scenes of devastation and carnage with corpses mixed with debris as far as the eye could see. Houses had been floated off their foundations and then flexed or collided with other buildings until they were destroyed. As many as 3,636 houses were destroyed, and almost 7,000 buildings were damaged or, basically, the entire city. There had been sixteen ships in the harbor when the storm struck, and all were damaged or destroyed.[28] All bridges to the mainland were washed out, leaving only railroads to the mainland and no way for the survivors to escape. More than 300 feet (91 m) of shoreline had been eroded away, leaving the ocean even closer to the city. The sand in the cemetery had been scoured away, leaving multiple coffins swept out to sea.

Surviving local authorities immediately declared martial law to prevent looting, and orders were issued to shoot anyone on sight who stole jewelry from the dead. The hot and humid Gulf coast weather returned as soon as the storm was gone, and the bodies began to decay and threaten the survivors with disease. Authorities quickly assembled teams of men to remove the bodies to dispose of them. About fifty African American survivors were recruited at gunpoint to serve on the death gangs. They collected as many dead bodies as they could and stacked them on carts, wheeling them down the few open streets to the beach. The workers were issued camphor-soaked handkerchiefs as face covering to help them endure the overwhelming smell and whiskey to help them deal with the horrors they observed. They loaded about 700 corpses onto barges on Monday afternoon, floated the barges into the gulf, and sank them with rocks. Unfortunately, this effort failed, and bodies began washing up onto the beach by that evening. The death gangs were immediately ordered to collect scrap lumber and build pyres to burn the bodies. The resulting fires burned day and night for weeks, filling the skies with foul-smelling smoke.[29]

When the reports and photos of the disaster reached the American public, they were horrified. Donations were sent immediately from all over the United States from children to millionaires and exceeded $1.25 million in total. The 78-year-old founder of the American Red Cross, Clara Barton, arrived with a team of relief workers on September 17.[30] Even Thomas Edison sent an assistant to Galveston to make the first film of a disaster on September 22. The film is still archived by the US Library of Congress.

The number of people killed in this storm is only approximate because so many victims were washed out to sea. Between 6,000 and 8,000 people lost their lives in the city of Galveston, but as many as 10,000–12,000 people were killed across the area. The official death toll is 8,000, making it the deadliest natural disaster in American history. The storm also left 10,000 people homeless. The total cost of damage done by the storm in Galveston, in the Caribbean, and across the United States and Eastern Canada was about $35.4 million, but the damaged in Galveston alone is estimated to be $30 million.

There were several significant changes that resulted from the storm. The Galveston residents expressed absolute intention to rebuild the city. In addition to rebuilding the houses, roads, bridges, and buildings, in 1902, they approved a bond issue to build a seawall on the shore facing the Gulf of Mexico. The seawall was built in 50-foot-long (15.2 m) sections beginning in 1902. Piling was driven up to 50 feet (15.2 m) deep, four feet (1.2 m) apart, and protected from undermining by sheet piling that extended 26 feet (7.9 m) into the sea floor. Reinforced concrete was applied to the pilings. The initial wall was 17 feet (5.2 m) high and 15 feet (4.6 m) thick at the base, 5 feet (1.5 m) thick at the top, and 3.5 miles (5.6 km) long. An additional section of the wall was built by the US government between December 1904 and October 1905 to protect the Fort Crockett Military Reservation. However, another major hurricane struck Galveston in 1915, and the storm surge topped the seawall, flooding the city and killing fifty-three people. In response, the height of the seawall was raised. The seawall was extended and modified numerous times in 1918–21, 1923, 1926, 1927, and 1950, increasing the current length to 10.4 miles (16.7 km).

The city of Galveston also decided to raise the elevation of the entire city. Sand was dredged between the jetties at the Galveston harbor entrance, deepening the channel into the harbor. The fill was used to raise the land surface behind the seawall to 16.5 feet (5 m), which supports the seawall. In the city, thousands of buildings, streetcar tracks, and pipes were elevated while the sand was pumped underneath them. Residents walked on catwalks up to 10 feet (3 m) above the ground surface in some areas. A church weighing at least 2,976 tons (2,700 mT) was raised 4.9 feet (1.5 m) with jacks in the modification. In addition, a low slope was also added west to Galveston Bay, draining the city streets into the bay during storms. This massive modification of more than 500 city blocks was completed in 1910 and required 16.3 million cubic yards (12.5 million m^3) of sand.

The Galveston hurricane also compelled the US Weather Bureau to establish a department to study hurricanes and to track and report on all Atlantic tropical storms and hurricanes. They developed communication channels both nationally and internationally to transmit these reports. Because of these changes, information about severe storms was broadly disseminated, and warnings were issued if human settlements were threatened.[31]

The real change in history, however, was how it affected the development of the Texas coast and cities, as well as the political and economic influence of the western Gulf of Mexico. Although the tenacious and industrious Galveston residents rebuilt the city quickly and installed safety measures to prevent another disaster, the storm had already caused irreparable damage to its image. Galveston had already begun to lose its dominance in terms of population by the time the storm struck. However, the horrifying pictures and stories dissuaded others from relocating there, further dropping it in the ranks of the most populous. Many of the millionaires who did not perish in the storm decided to relocate to safer areas. Companies also slowly but steadily chose to relocate to the safer and more populous areas. With time, Houston grew as the industrial powerhouse of the region as Galveston lost all of its power and was relegated to the local vacation destination as it is today. The hurricane of 1915 was the final straw in this transition. Much of the port traffic that previously went to Galveston also shifted to New Orleans. These two cities control the economic and political power in the region and have done so for most of the twentieth century.

1970 Cyclone Bhola

The 1970 Bhola cyclone in East Pakistan was so impactful that it changed history in the area and the world in several major aspects. The area was already a powder keg for conflict with all of the major powers in the Cold War keenly interested in the situation. The relationship between East and West Pakistan had been contentious for a decade by the time the storm struck. It was under military rule during this time. The control of the government was in West Pakistan, with East Pakistan more like a province. The people of East Pakistan felt poorly served in terms of the distribution of the country's resources and services. In addition, two weeks prior to the Bhola cyclone on October 23, 1970, the area was threatened by another cyclone, and a large area of East Pakistan was evacuated at great expense to the Pakistani government. That storm weakened and veered away from the coast, making the effort a waste. For this reason, the Pakistani government was reluctant to make the same mistake again.

The Bhola cyclone began as the remnant of a tropical storm that drifted west across Malaya on November 5, 1970. The system evolved into a tropical depression on November 8 over the central Bay of Bengal.[32] The depression then strengthened to tropical storm status on November 9. The storm steadily strengthened to cyclone status while accelerating northward. It was then that Pakistan began monitoring it as a potential threat. The winds continued to accelerate, reaching 115 miles per hour (185 kph) by November 11 when it swung sharply northeastward and directly toward East Pakistan. The cyclone impacted the coast on November 12 with average wind speeds of 130 miles per hour (205 kph) and gusts of 150 miles per hour (240 kph).[33]

East Pakistan is considered to be one of the most disaster-prone areas in the world. It has experienced numerous very deadly cyclones both before and after Bhola. Hazards threaten about 98 percent of the population and more than 97 percent of the land area. It sits in the Ganges River delta with one-third of the land as coastal shoreline. Approximately 80 percent of the land is classified as floodplain, and nearly 67 percent is less than 16 feet (5 m) in elevation. At least 20 percent of the country is flooded during annual river floods. It is the most vulnerable to cyclonic storms with about 5–6 percent of the global total. However, it also accounts for 80–90 percent of the global property loss and human death tolls.[34]

The now category-4 cyclone struck the East Pakistan coast during local high tide on the evening of November 12.[35] It made landfall in the south-central part of the country with winds measured at 139 miles per hour (224 kph) with, more importantly, a storm surge of 35 feet (10.6 m) high. Because it was high tide, the sea level was already elevated. At the time, there was no widespread warning system in place, so the residents had no warning much less an evacuation being ordered. They were totally unaware and vulnerable. The huge storm surge completely submerged whole islands and coastal areas with only the tops of trees sticking out of the flood waters. Residents who weren't swept away climbed trees and hung onto the trunks through the rain and battering winds. Government reports indicate that most of the dead were women and children because they were unable to hold onto the trees through the storm. The cyclone quickly degraded to a tropical storm once it made landfall.[36]

The 1970 Bhola cyclone was perhaps the worst natural disaster of the twentieth century. Officially, there were 200,000 confirmed burials, 50,000–100,000 people missing, and presumed dead for a total of about 300,000 deaths. However, some reports place the death toll as high as 500,000, which officially, makes it the deadliest tropical cyclone in history and possibly the worst natural disaster ever. So many victims were washed out to sea that the actual totals will never be known. The damage to the area was apocalyptic. The areas of highest impact were Bhola, Chattogram, Cox's Bazar, Bagerhat, and Noakhali. Whole fishing villages disappeared. Across the impacted area, at least 85 percent of homes were destroyed or badly damaged. At least 90 percent of ocean-going fishermen suffered severe losses, including the sinking of 9,000 ocean fishing boats. At least 46,000 of the 77,000 onshore fishermen were killed, and 40 percent of the survivors were severely impacted. In an area where 80 percent of the protein is from fish, at least 65 percent of the fishing capacity was destroyed. Damage to agriculture was also catastrophic with the destruction of some $63 million (US) worth of crops covering one million acres, 280,000 cattle, and a half million poultry. The total estimated damage was $86 million (US) in 1970 dollars.

Ghastly images of the catastrophe appeared in the evening news worldwide. Bloated human corpses and cattle carcasses floating in flooded areas were seen rotting in the heat and humidity. Locally, survivors could be seen marooned on small islands, waving at the passing helicopters. Aid poured into Pakistan from around the world. President Nixon committed $10 million (US) to disaster relief, and the World Bank offered a reconstruction plan $185 million (US) to Pakistan. The government leaders in West Pakistan, however, were slow to institute inadequate disaster relief to the impacted areas, which ignited anger among the people of East Pakistan. They believed that West Pakistan was neglectful of their needs in general, but to be neglectful during a disaster was inexcusable. The government was strongly condemned by many countries and internal political factions including the leader of East Pakistan's dominant political party, the Awami League.[37] Waterborne disease swept through the area, and offers of aid from nearby India were declined. Four months after the disaster, 600,000 people in East Pakistan still lacked adequate shelter.

Despite the utter chaos in East Pakistan, country-wide national and provincial elections took place on December 7, 1970. This was the first popular election since Pakistan declared independence in 1947. Many opposition parties boycotted the elections, and the Awami League won 160 seats in the National Assembly and the vast majority of the Provincial Assembly seats.

These election results meant that the Awami League would lead the government and draft the country's first democratic constitution.

The head of the military who formerly controlled Pakistan was not pleased with these results. On January 3, 1971, two million people rallied in the East Pakistan capital, and the leader called for full provincial autonomy and a separate militia. The people of East Pakistan felt strongly that West Pakistan was unconcerned with their security. Feeling this threat, the military leader postponed the first session of the new National Assembly, scheduled for March 3, 1971. This resulted in violent demonstrations, including clashes between protestors and armed police and military forces. In mid-March, the military leaders met with Awami League leaders to broker a deal, but no agreement could be reached. In response, more military troops were secretly flown into East Pakistan. Finally, on March 23, the Awami League proclaimed provincial autonomy and warned that if it was not recognized within forty-eight hours, East Pakistan would erupt in violence. By then, the people of East Pakistan wanted independence, not autonomy.[38]

On March 25, Pakistan's military leader departed East Pakistan and instructed the troops to launch a full assault on East Pakistan to reassert West Pakistan's dominance over them. Martial law was declared, and tanks and armored vehicles invaded the East Pakistan capital city, crushing and killing everyone in their path. The university was attacked with students and professors being executed. The general in charge of the assault became known as the "Butcher of Bengal" as a result. On March 26, Awami League leader declared the independence of Bangladesh, and he was promptly arrested and accused of treason. The military regime banned all political activities, and open rebellion broke out in East Pakistan.

Meanwhile, on March 27, India's prime minister Gandhi declared verbal support for East Pakistan in an open session. On March 31, the Indian Parliament adopted a resolution of verbal support. News of the Pakistani military attacking unarmed civilians led to overwhelming support for East

Pakistan independence in the border region of West Bengal, India. They collected money, food, and medical supplies for those fleeing the violence by entering India. The Indian government began allowing large numbers of Pakistani refugees into the country. The Indian Army began to covertly train Pakistani freedom fighters who had crossed the border. Some Indians from West Bengal and Tripura joined the group. On April 10, 1971, a provisional Bangladesh government proclaimed independence and seven days later set up headquarters in Calcutta, West Bengal, India. As the situation worsened in East Pakistan, the Pakistani military was being accused of genocide on an international basis. By then, millions of refugees were crossing into India, and there were fears of cholera outbreaks in the numerous congested refugee camps. By May 1971, three million Bengali refugees had crossed into India, and thousands arrived daily.

The superpowers including the United States, the USSR, and China began to take less than peripheral roles in the situation. The United States was allied with the West Pakistan government and turned a blind eye to the genocide. They even diverted a nuclear submarine from monitoring and assisting in the Vietnam War to the Bay of Bengal to threaten India into inaction in the conflict. India was afraid that if they aided Bangladesh, China would seize the opportunity and attack them along their northern border as had happened in the past. The United States vowed not to help India if they incurred Chinese aggression. As such, India first negotiated a treaty with the Soviet Union to gain some significant support if they chose to act. They later negotiated a nonaggression agreement with China over the conflict, leaving only the United States outright opposing them. This series of interactions established the political arrangement for the eighteen-year remainder of the Cold War in the Far East.

With the major political barriers removed, a thirteen-day war began on December 3, 1971, between India and Pakistan. The Pakistani Air Force attacked numerous airfields in north and west India from West Pakistan, and their infantry shelled numerous Indian military posts to start the conflict. The Indian government conducted a major retaliation which began the Bangladesh Liberation War or the third India–Pakistan War. The United States sent warships to the region but did not participate in the conflict. The war for Bangladesh was quickly resolved with the surrender of Pakistan on December 16, 1971. However, the situation grew tense for a time, and there was worry that it would precipitate the Third World War with so many major powers involved. The incident also severely damaged US–India relations, generating mistrust of the United States by the Indian media, the public, and the Gandhi government and drove India to switch their alliance to the Soviet Union. As a result, India increased diplomatic relations with North Vietnam in January 1972. Because of the improved US–China relations, China vetoed the entry of Bangladesh into the United Nations in August 1972.

Another significant historical change resulting from the Bhola cyclone disaster was cultural and positive. George Harrison, lead guitarist for the recently disbanded Beatles, had long been interested in Indian culture, infusing the local music into his songs. He teamed with musician Ravi Shankar from West Bengal, India, the area that had been so crucial in the conflict, to organize the first major benefit concert for a social cause ever. The resulting spectacular Concert for Bangladesh was held on Sunday, August 1, 1971, in Madison Square Garden, New York City, between 2:30 and 8:00 p.m. The performers featured rock superstars such as Bob Dylan, Eric Clapton, Ringo Starr, Leon Russell, and the band Badfinger. The effort was supremely successful and resulted in a motion picture and album released the following year. This concert introduced the phenomenon of concerts to raise money for social causes such as LiveAid, FarmAid, and many others. Concert for Bangladesh also raised more than $12 million for relief efforts in Bangladesh.[39]

The Bhola cyclone may have been the most catastrophic cyclone, but it certainly was not the first or last. Bangladesh prioritized cyclone readiness in the new nation. They developed a plan for early warning systems, cyclone preparedness programs, public education about cyclones, construction of cyclone shelters, and forestation programs. However, the country did not have the funding to undertake all of these efforts quickly. Unfortunately, the efforts were not quick enough. On April 29, 1991, one of the most powerful cyclones in the Indian Ocean struck Bangladesh with winds of up to 145 miles per hour (235 kph) and a storm surge of up to 20 feet (6.1 m). The coast was again submerged, killing more than 138,866 people and causing $1.7 billion (US) in damage. This greatly accelerated the cyclone preparedness in Bangladesh. By the time the next major cyclone struck on November 15, 2007, they were better prepared. The category 4 cyclone had wind speeds of 150 miles per hour (240 kph) and a storm surge of up to 20 feet (6 m) high. However, even though the storm was as strong as the previous disasters, the death toll was a much reduced 4,407 people, 100 times less than Bhola. There have been other cyclones since 2007, but none as large as these three. All have had death tolls of 150 people or less. This shows the effectiveness of the cyclone reduction efforts in Bangladesh.[40]

6

Tornadoes

High travel speed and extremely fast winds make tornadoes the most ferocious storms on earth. Before modern radar and technology, they would appear with no warning and demolish entire towns in minutes. Fortunately, tornadoes are small, short-lived, and destructive tornadoes are largely restricted to specific locations. The relatively small area of impact makes it difficult for tornadoes to change history more than locally. Outbreaks of multiple tornadoes can do more widespread damage, but it is still quite localized around each one.

Tornadoes form in active weather systems that include thunderstorms. Hot, moist air near the ground interacting with cooler air above produces vigorous vertical circulation to high altitudes. The air circulation forms convection cells as individual in single-cell thunderstorms, multicell with multiple single cells, or supercell when the circulation is especially intense. Tornadoes form in multicell storms but, by far, the primary tornado producers are supercell storms. The dominant circulation direction of the storm controls its characteristics. Storms dominated by upward flow carry warm, moist air to higher and cooler levels, which causes excessive condensation. These storms yield heavy precipitation but little wind. Storms with strong downward flow produce damaging straight-line winds and microbursts but little precipitation. Supercell storms are characterized by intense upward and downward flow. The upward flow causes storms to grow to great height with a characteristic anvil shape of thick, cumulus clouds and a flat top that extends behind the storm.

The intense upward and downward circulation can produce hail. This same motion can also produce a tornado. This is the reason that hail is commonly associated with tornadoes.[1]

Storms that produce tornadoes begin with horizontal rolling motion. Moving cool, dry air at high altitude rides over warm, moist air that moves in another direction. The cool air is dense and sinks as the warm, less dense air rises which creates a rolling air in the unstable atmosphere. The instability in vertical flow flips the developing storm on end to produce a spinning cloud. These storms primarily spin counterclockwise, but they can also spin clockwise. The spinning cloud is a mesocyclone, and the upward flow makes it grow taller and stronger, but the downward motion generates a tornado. The first indication of downward motion is bulbous, mammatus clouds, followed by the development of a wall cloud below the spinning storm cloud. A funnel cloud drops downward from the center of the spinning wall cloud and becomes a tornado if it reaches the ground. The tornado spins very fast, producing ferocious winds. In supercells with tornadoes, one part of the storm is dominated by updraft and another by downdraft. The updraft area has hail and heavy rain, but the downdraft can contain a tornado.

Tornado strength classification is based on wind speed. The Fujita scale was the accepted classification until 2007. It ranged from F0 to F5 as follows:

Rating	**Name**	**Wind speed MPH(KPH)**	**Frequency**
F0	gale tornado	40–72 (64–116)	29%
F1	moderate tornado	73–112 (117–180)	40%
F2	significant tornado	113–157 (181–252)	24%
F3	severe tornado	158–206 (253–330)	6%
F4	devastating tornado	207–260 (331–417)	1–2%
F5	incredible tornado	261–318 (418–509)	<<1%
F6	inconceivable tornado	319–379 (510–606)	Ever?

With usage, the Fujita scale was found to have several weaknesses. It is subjective because it is based only on damage caused by a tornado. The problem is that damage is not distinguished among the different types of construction. Also, it is difficult to apply if certain damage indicators are absent. If a tornado does not strike any buildings, a Fujita scale rating cannot be designated. Rating is based on the opinion of observers and only on the worst damage, even if only one building shows damage. Additionally, it overestimates wind speeds in tornadoes greater than F3.

For these and other reasons, on February 1, 2007, the Fujita scale was replaced with the Enhanced Fujita (EF) scale in the United States. The EF-scale tornado rating is based on twenty-eight damage indicators. Each indicator includes a description of the construction, and the degree of damage is determined by each indicator present. Wind speed is estimated by comparison of the observed damage with a calibrated guide. The observed damage constrains a wind speed range. The damage indicators include many types of buildings, towers, and trees. Final wind speed is estimated using as many features as possible. The EF system is a more inclusive and accurate system than the Fujita scale. The wind speed is estimated as lasting for three seconds or more rather than just peak speed. The scale is

EF-0: 65–85 mph (104–137 kph).

EF-1: 86–110 mph (138–177 kph).

EF-2: 111–135 mph (178–217 kph).

EF-3: 136–165 mph (218–264 kph).

EF-4: 166–200 mph (265–320 kph).

EF-5: >200 mph (320 kph).

Most funnel clouds do not reach the ground and become tornadoes. If they do, tornadoes begin as EF-0 storms and lift and touch down a few times as they start. Once on the ground, they can intensify to higher ratings. They usually start as thin funnels but intensify to a wider funnel. The tornado path length and direction can vary. Some tornadoes travel in straight paths, but others sweep side to side with longer funnels tending to sweep more. The forward speed of tornadoes ranges from 20 to 60 miles per hour (32–96 kph), depending on the situation. Most tornadoes stay on the ground for only a few seconds to a few minutes, especially for storms that are EF0 or EF1. However, larger storms can remain on the ground for much longer periods. They can be on the ground up to three hours in the worst cases.[2]

An indication that a tornado has reached the ground is a debris cloud at the base. Tornadoes can spread tons of debris in all directions. The debris cloud includes fragments of construction materials, vegetation, and soil. The powerful winds pulverize much material and produce dust or particulate that leaves a trail along the tornado path. The dust floats in the air for days around the affected area.

Tri-State Tornado

The massive Tri-State tornado was the deadliest and most destructive tornado in the history of the United States. It left a path of destruction across Missouri, Illinois, and Indiana on Wednesday, March 18, 1925, as the most powerful of a tornado outbreak that included at least twelve other strong tornadoes. The path of the huge tornado was an amazing 151–235 miles (243–378 km) long, extending from southeastern Missouri to southwestern Indiana. This is the longest tornado path ever recorded. It left at least 695 people dead, making it also the deadliest tornado in the history of United States. Only Bangladesh has experienced deadlier tornadoes.[3,4]

The weather situation that caused the Tri-State tornado was no different than that which causes most severe weather in the midwestern United States. A Canadian cold air mass swept in from the northwest into the United States with a low-pressure system along the cold front. The low swept into Montana on March 16, 1925. On March 17, a fast-moving warm front led the northward incursion of warm, moist air from the Gulf of Mexico into the southern United States.[5] This early warm air intensified the cold front by increasing the contrast across the front. The two fronts met at the surface low-pressure system that moved into northeastern Oklahoma by 7:00 a.m. on the morning of March 18. The low continuously strengthened, drawing in Gulf Coast air as it moved northeast into Missouri. Several other smaller low-pressure systems began forming along the cold front, all contributing to a very dangerous situation. As a result, early morning showers and thunderstorms were widespread across Kansas to Indiana, but the situation began to stabilize as the morning wore on.[6]

By midday, however, the main low-pressure system had deepened into a large supercell. The temperature had reached a warm 60–70 °F (16–21 °C) in Missouri and Illinois, and tornadoes were reported in the area by 12:40 a.m.[7] The Tri-State wedge tornado was first seen about 3 miles (4.8 km) north-northwest of Ellington, Missouri, at 1:01 p.m. where it killed a farmer. It appears to have formed directly under the low-pressure system from a large supercell. It was not larger than other Midwestern supercells, and the low-pressure system was no deeper than most others, but they produced the deadliest and perhaps the longest-lasting and most powerful tornado in history. It would devastate parts of Missouri, Illinois, and Indiana, crossing thirteen counties including Reynolds, Iron, Madison, Bollinger, and Perry in Missouri, Jackson, Williamson, Franklin, Hamilton, and White in Illinois, and Posey, Gibson, and Pike in Indiana. It would stay on the ground for more than 3.5 hours and devastate more than nineteen communities, some would be abandoned as a result of the damage.

Once formed, the Tri-State tornado crossed farmland, rural areas, towns, and cities as it traveled generally northeast. It struck the mining town of Annapolis, Missouri, and the small town of Leadanna as it steadily grew stronger. About 90 percent of Annapolis was destroyed. In Bollinger County where thirty-two children were injured when their school was struck, deep scouring of the ground surface was reported, and large iron sheets were lifted off the ground and carried as far as 50 miles (80 km) away. It killed two people and injured seventy-five there.[8,9] The tornado then developed a double funnel as it strengthened toward the Mississippi River. At least eleven to thirteen people were killed in Missouri by the Tri-State tornado, and as many as 200 were injured.

When the now powerful Tri-State tornado crossed the Mississippi River and into southern Illinois, it grew even stronger and became rain wrapped or high precipitation, which meant it could not be seen until it was on top of the victims. In places where the rain subsided, the funnel was wrapped in a huge ball of dust and debris and also could not be seen until it was too late. As it passed through rural areas, it left deep scour marks in the ground. Then at 2:30 p.m., it obliterated the entire small town of Gorham, Illinois. It destroyed 75 percent of the buildings and damaged the rest.[10] The railroad tracks were even torn up by the powerful wind. At least thirty-six people were killed and 170 were injured, accounting for more than half the population of Gorham.

The Tri-State tornado reached its peak intensity in Gorham and as it sped northeast. It is calculated that the forward speed averaged about 62 miles per hour (100 kph), but it achieved speeds of 73 miles per hour (117 kph) locally. This is extremely fast for tornadoes and would have given victims virtually no time to seek shelter. During this segment, the tornado had a width of almost 1 mile (1.6 km). Some claim that the tornado achieved a wind speed of 300 miles per hour (483 kph) at this point, but all agree that it was an EF5 (F5)[11] tornado. This awesome storm slammed into Murphysboro, Illinois, with catastrophic results. It destroyed or severely damaged most buildings in the city, completely flattening entire neighborhoods. Several schools were severely damaged, killing some twenty-six children. In all, 234 people lost their lives which constitutes the greatest loss of life from a tornado in a single city in the history of the United States. There were also 623 people injured and caused at least $10 million (US) in damage.[12]

The next community struck by the Tri-State tornado was the village of De Soto. It too was obliterated with rows of houses completely removed from their foundations. In all, fifty-six people were killed, and at least, 105 were injured. The De Soto School partially collapsed with brick walls falling in and crushing students. This accounted for the deaths of thirty-three students, making it the worst death toll at a school from a tornado in the history of the United States. A 300-foot (91 m) section of railroad track was torn up by the powerful wind.

After destroying part of the town of Bush and killing ten people, the tornado plowed through the center of the town of West Frankfort. At this point, it was a high-precipitation, rain, and hail-wrapped tornado, making it invisible to the residents and victims in the town. Survivors report that the storm appeared to be a nebulous, boiling mass of clouds that rolled along through the town. In all, the tornado killed 102 people in West Frankfort, and 389 were injured. Then, the mining village of Caldwell was destroyed killing twenty-four people. This was followed by the town of Parrish being flattened where thirty-three people died, and sixty were injured. The destruction of Parrish was so overwhelming and so many survivors and businesses relocated to another area that the town was abandoned. In Franklin County, Illinois, the death toll was 192. The tornado entered the rural Hamilton and White Counties, where it widened to a huge 1.5 miles (2.4 km), which was the widest of the entire path. At least 65 people were also killed in this area, 45 right away and another 20 of the 120 who were injured within a few days.

The Tri-State tornado then crossed into its third state, Indiana, where it was still strong enough to destroy everything in its path. It first demolished the town of Griffin where reportedly every building in the town was either damaged or destroyed including 150 houses. The death toll in Griffin and the surrounding areas was forty-six people, and there were 202 injuries. After the tornado killed nine people in Owensville, it destroyed the south side of the town of Princeton and killed forty-four people. At least 152 people were injured. The tornado then traveled at least 10 miles (16 km) northeastward into Pike County before dissipating at about 4:38 p.m. At least ninety-nine people were killed in Indiana.

In all, the 1925 Tri-State tornado killed at least 695 people, including 588 in Illinois, ninety-five in Indiana, and twelve in Missouri. Some studies report that deaths most likely exceeded 747. The calculated number of people injured was 2,298, at a minimum. Four of the impacted communities were abandoned because the damage and death were so devastating that residents chose to relocate. Some of the rural areas never fully recovered. More than 15,000 houses were destroyed by the tornado, and the total damage was estimated to be $16.5 million in 1925 US dollars, and at least $1.4 billion in 1997 US dollars.[13]

The extremely long path was about 235 miles (378 km) and the tornado was on the ground from around 1:01 p.m. to about 4:38 p.m. This is an amazing 3:37. The extremely long path length was found to contain thirty-two gaps, suggesting that the Tri-State tornado may have actually been multiple tornadoes, although the study was completed many decades after the event. Certainly, multiple vortices were observed at several locations, and the main tornado could have ended and a subsidiary tornado picked up. However, it was concluded that there are only minor gaps for a distance of 174 miles (280 km) from central Madison County, Missouri, to Pike County, Indiana, and that it was certainly continuous for that long segment. Therefore, although it cannot be ruled out that Tri-State tornado was actually several tornadoes, it was at least primarily a single tornado, if not completely a single tornado.[14]

Reconstructions of the meteorological conditions that spawned the killer tornado were conducted, so another Tri-State tornado could be identified and predicted ahead of time to reduce the death toll. It was concluded that the supercell that produced the Tri-State tornado was likely isolated from other storms throughout its path, which argues for a single tornado rather than multiple tornadoes. However, there was no specific feature in the conditions that could explain why it was so extreme. It was concluded that the supercell moved rapidly eastward right along the front, which kept it in a very favorable tornado environment for several hours. Normally, the supercells are so dynamic with regard to updrafts and downdrafts that the conditions quickly become unfavorable for a single tornado. Therefore, the reason that the Tri-State tornado became so powerful and lasted so long is that things stayed the same for a long time.

As for changes to history, most were local. Besides the four towns that were abandoned, the residents of the remaining towns and rural areas were traumatized by the tornado. Immediately after rescue and relief operations concluded, there was a rash of hundreds of storm shelters being constructed. Many people in rural areas simply dug holes in the ground large enough to hold their families to protect them from the falling walls and airborne debris in a tornado. For several decades after the Tri-State tornado, residents of Murphysboro and West Frankfort were so frightened that they rushed to storm shelters when even just a dark storm cloud appeared. Another result of the Tri-State tornado was a major change in property insurance coverage in the Midwest. In 1930, the insurance industry added tornado damage to property coverage in Midwestern policies.[15]

In 1925, the word tornado was forbidden in weather forecasts because they could not be predicted, and mentioning them would cause public panic. Stories and photos of the destruction from the 1925 Tri-State tornado covered newspapers worldwide and captivated the attention of Americans. They became a priority to many, and discussions bantered across the press for weeks after the storm. It was this tornado that indelibly associated tornadoes with the Midwest in the minds of Americans. The main focus of the debate was tornado prediction, and the premise that there must be some way to predict them or possibly build structures to withstand them. Determining ways to predict tornadoes became a priority after the Tri-State tornado.[16] Was the American fascination with tornadoes caused by the Tri-State tornado part of the impetus to film the Wizard of Oz movie just twelve years later?

The first successful prediction of a tornado would not occur for twenty-three years after the Tri-State tornado. On March 20, 1948, a tornado struck the Tinker Air Force Base, Oklahoma, and the surrounding area without warning. The storm caused more than $10 million (US) in damage as it crossed the airfield and overturned aircraft, toppled power lines, and broke the control tower windows. The two air force weather officers on duty kept careful notes on the conditions that produced the tornado. The two were again on duty 5 days later on the afternoon of March 25 when they observed similar atmospheric conditions. They contacted the base Commanding General with their observations, and he issued the first-ever tornado forecast. As predicted, a tornado struck Tinker Air Force Base that evening, but all equipment and planes had been secured, reducing the impact.[17] By 1950, the restriction of not predicting tornadoes in weather forecasts had been lifted.

1974 Super Outbreak

Strong tornadoes commonly occur in multiples. For example, the Tri-State tornado was one of more than a dozen that were produced by the weather system and, therefore, was part of an outbreak. When there are more than a few strong tornadoes from two or more distinct supercells, it is a tornado outbreak. The intensity of outbreaks is ranked using a point system called an outbreak intensity score (OIS). In this system, each EF2 tornado in an outbreak is awarded 2 points, whereas each EF3 gets 5 points, each EF4 gets 10 points, and each EF5 gets 15 points. The total points for the outbreak then determines the OIS, where 2–6 points is a weak outbreak, 7–10 is minor, 11–29 is significant, 30–79 is major, 80–119 is devastating, 120–249 is historic, and 250 or more is a super outbreak. Using this system, the 1974 Super Outbreak was the most powerful outbreak in recorded history at 446 points despite the 2011 Super

Outbreak having had 360 tornadoes spread over five days compared to the 1974 outbreak at 148 tornadoes in one day.[18]

The 1974 Super Outbreak of tornadoes was the first documented outbreak to produce more than 100 tornadoes in less than twenty-four hours and remains the second most numerous ever. Between April 3 and 4, 1974, 148 tornadoes were confirmed in thirteen states and Ontario, Canada. In terms of the power of the storms, 1974 remains the most violent with thirty tornadoes of F4 and F5. The storms struck an area from the Gulf Coast to Canada including Alabama, Georgia, Illinois, Indiana, Kentucky, Michigan, Mississippi, New York, North Carolina, Ohio, Tennessee, Virginia, and West Virginia. Although this event had a minor impact on world history, numerous changes to the system of how tornadoes were forecast, studied, and announced to the public were made over the next few decades as a result. These changes saved hundreds to thousands of lives.[19,20]

On April 1, 1974, a strong low-pressure system formed in the Central Plains of the United States. It moved into the Mississippi and Ohio River Valleys over the next few days. A large cold air mass also moved eastward, and it collided with a surge of warm, moist air from the Gulf of Mexico, forming a strong cold front. This was coupled with a mid-level jet stream with winds of 120 miles per hour (200 kph), producing severe instability. This produced a few F2 and F3 tornadoes in parts of the Ohio River Valley and Southern United States on April 1 and 2. By early in the day on April 3, the National Weather Service forecast severe weather. There was a large low-pressure trough across the central United States, with rotation around the base. The front developed three squall lines that generated the severe weather with the second line producing the most powerful storms.[21]

The situation was so unstable that by 11:30 a.m., an F3 tornado was reported in Tennessee. By 12:20 p.m., baseball-sized hail was reported in Illinois. Three more supercells were producing tornadoes in the Tennessee and Ohio River Valleys by 1:00 p.m., and they were moving at speeds up to 69 miles per hour

(111 kph) in a northeasterly direction. This severe weather sped into Illinois, Indiana, and northern Kentucky where the second squall line produced three major supercells and several smaller ones. Between 2:20 and 3:20 p.m., the large supercells developed that would later produce deadly F5 tornadoes. This middle squall line extended from Pennsylvania to Tennessee. The first F5 tornado formed at about 3:20 p.m. in Depauw, Indiana, where it traveled for 65 miles (105 km). Fortunately, it was mostly in rural areas and killed only six people while injuring eighty-six. The second F5 tornado formed at 3:25 p.m. and traveled about 32 miles (51 km).[22] It completely destroyed the town of Brandenburg, Kentucky, stripping houses from their foundations and even stripping up the grass from lawns. In all, thirty-one people were killed by this tornado, and 257 were left injured. This still ranks as the most powerful tornado to have ever struck Kentucky.

The third tornado of this group was the deadliest. Around 4:30 p.m., a multivortex tornado formed to the southeast of Xenia, Ohio. This F5 and nearly F6 tornado struck the city traveling at 50 miles per hour (80 kph), carving a 0.5-mile (0.8 km) wide path. It flattened a large part of the city. More than 1,400 buildings were heavily damaged or destroyed and had to be demolished. It was responsible for thirty-two deaths and about 1,150 injuries. Damage of Xenia alone is estimated at $471.7 million in 2013 dollars.[23] The only city that President Richard Nixon visited, of the many cities damaged in the outbreak, was Xenia, Ohio, and he declared it to be the worst disaster he had ever seen. He provided federal aid to Xenia, and most of the city was completely rebuilt. It took ten months of research for Dr. Ted Fujita to decide that it was an F5 and not an F6 tornado.

Another tornado began shortly after 4:30 p.m. and was important because it was shown live on the evening news in Cincinnati, Ohio, possibly a first for tornado coverage. The Sayler Park tornado was an F5 storm that struck the city of Cincinnati. Fortunately, it had weakened by the time it reached the city and did most of its damage prior to that. In all, the tornado killed 3 people and injured 210.

Meanwhile, by 3:50 p.m., additional supercells developed in Illinois and Indiana, which produced additional tornadoes up to F3. One of these traveled 109 miles (180 km). By 6:00 p.m., the southern part of the second squall line struck Tennessee, Mississippi, and Alabama, producing some of the strongest tornadoes of the outbreak. In Alabama, there were two F5 tornadoes that each struck the town of Tanner about thirty minutes apart. The first tornado formed at about 6:30 p.m. and was on the ground for about ninety minutes. This 3/4-mile (1.2 km)-wide tornado caused extensive damage, killing twenty-eight people and injuring 267. The second F5 tornado to strike Tanner touched down at 7:35 p.m.. It cut a path 1500 feet (457 m) wide and 83 miles (134 km) long. It also did extensive damage. Fortunately, many residents were still taking shelter from the first tornado, reducing the death toll to sixteen people with 190 injuries. An extremely powerful F5 tornado destroyed the town of Guin, Alabama, and was the longest-duration F5 tornado of outbreak. The tornado touched down at 8:50 p.m. and traveled 79.5 miles (128 km) before dissipating at 10:30 p.m. The tornado grew extremely violent as it struck Gwen, causing catastrophic damage and killing 28 people while injuring 272.

However, F5 tornadoes were not the only deadly ones. F4 tornadoes could be equally deadly especially if they occurred in tornado families. Clusters of supercells along the same squall line that were spatially and meteorologically related formed tornado families. The Monticello tornado family was the deadliest of the 1974 Super Outbreak. The family consisted of at least eight tornadoes that tracked from Illinois to Michigan from about 2:45 p.m. to 7:00 p.m. in a path 260 miles (418 km) long. The largest of these tornadoes was the F4 Monticello tornado, which tracked 121 miles (195 km). Although this tornado destroyed several towns, the most impressive damage was done in Monticello, Indiana, where it caused $100 million in damage. Of the 20 deaths and 379 injuries caused by the Monticello family, 18 of the deaths and 362 of the injuries were caused by the Monticello tornado alone.[24]

The third squall line also became active beginning at about 4:00 p.m. This line was located to the West of the other squall lines. It produced tornadoes up to F3, which also did extensive damage. Storms produced strong straight-line winds and hail, which did extensive damage to St Louis, Missouri. One F3 tornado in this line was on the ground for 121 miles (195 km).

Other areas were also struck by tornadoes and other severe weather lasted well into April 4th. Michigan was struck by an overnight snowstorm following the weak tornadoes during the day. Other weak and late tornadoes struck Ontario and western New York. By the early morning of April 4, the southern parts of the squall line's activity moved through the Appalachian Mountains and produced weaker tornadoes across the southeastern United States. The final tornado of the 1974 Super Outbreak dissipated in Caldwell County, North Carolina, at 7:00 a.m.

The 1974 Super Outbreak produced the most vigorous tornadic activity in recorded history. Over a period of just eighteen hours, it produced 148 tornadoes from 30 powerful supercells. It produced seven F5 tornadoes and twenty-three F4 tornadoes associated with thirty-three distinct tornado families. All combined, the tornadoes had a path length of more than 2,600 miles (4,184 km) and damaged or destroyed an area of more than 900 square miles (2,331 km^2). At one point, there were sixteen tornadoes on the ground at the same time. Ten of the most impacted states were declared as federal disaster areas. The estimated cost of the damage was $843 million, but in 2024 dollars, the damage was $5.3 billion. This event led to the deaths of 335 people, and more than 6,000 people were injured.[25]

The 1974 Super Outbreak led to several changes in the United States both local and at the federal level. Americans learned of this disaster through the horrifying pictures, videos, and stories in the media on a daily basis for the next few weeks. Public concern quickly put pressure on elected officials, and by as early as April 10, the Senate had passed the 1974 Disaster Relief Act. This bill streamlined declaring a federal disaster area from any event and sped up the awarding of disaster relief aid, as well as creating a blueprint for a disaster relief agency. This bill was signed into law on May 22 by President Nixon.[26]

The 1974 Super Outbreak overwhelmed the National Weather Service offices and early warning systems across most states that were impacted. At one point, there were so many tornadoes on the ground and favorable conditions for developing tornadoes in Indiana that the National Weather Service issued a warning for the entire state. It was the only time that an entire state was under a tornado warning. At the time, there were only a few rudimentary weather radar systems that could barely discern tornadic activity. Meteorologists had to wait so long to confirm a tornado that there was little warning time for residents. As a result of this outbreak, the development of advanced weather radar was greatly accelerated and installed in many more areas. Warnings needed to be issued quicker as well. NOAA increased its warning radio coverage from 50 to 60 major cities to 330 cities of all sizes after the outbreak. Finally, because so many schools had been impacted by tornadoes, tornado drills and education were practiced at all schools. In addition, the number of tornado sirens was greatly increased across the Midwest.[27]

This was the first real test of the F system for ranking the strength of tornadoes, which was designed in 1971. It showed that the system needed modification. Dr. Ted Fujita, who devised the Fujita scale, and his research team performed a ten-month study on the 1974 Super Outbreak. His funding to research tornadoes had almost run out, but the outbreak was so impactful that it convinced the funding agencies to support his research for the rest of his career. The detailed studies of the tornadoes of the outbreak also debunked a number of tornado myths that persisted to that time.

Studies showed that buildings don't explode because of the pressure difference when a tornado passes over them. Before this, people wasted time opening windows to equalize pressure as tornadoes approached when they should have been sheltering. Another myth was that tornadoes never climbed steep hills or mountains, but the F5 Guin, Alabama, tornado climbed the 1640-ft (500 m) high Monte Sano Mountain. There was another myth that tornadoes weakened as they crossed rivers, but the powerful F5 Sayler Park, Ohio, tornado crossed the Ohio River without weakening. These new insights allowed residents to better prepare for and survive tornadoes.

1999 Moore Tornado

The 1999 Moore, Oklahoma, tornado was the most powerful tornado ever documented, possibly achieving F6 classification. It was also the deadliest in the United States in over twenty years. It changed the way the United States approached tornado preparation and protection as the next step after the 1974 Super Outbreak. In this way, it altered local and United States, if not world history. Our tornado warning and protection systems today largely reflect advances gained in response to these events.

The Moore tornado was one in the major two-day 1999 Oklahoma tornado outbreaks or the 1999 Great Plains tornado outbreak. This outbreak produced seventy-three tornadoes on May 3, 1999, across five states and an additional fifty-eight tornadoes on May 4–5 stretching south into Texas and east into the Tennessee Valley and the Southeast for a total of 131 tornadoes making it the third largest outbreak ever,[28,29] although some sources stretch it from May 2–7 and make it even bigger. In any event, more than sixty tornadoes struck Oklahoma on May 3, making it the largest tornado outbreak in Oklahoma history. The previous one-day record was twenty-six tornadoes.[30]

The meteorology of the situation was not that unusual with a large warm, moist air mass moving north from the Gulf of Mexico and being met along a cold front by a cold, dry air mass sweeping in from the northwest. However, the jet stream had swung far south into Texas and approached the front at a high angle from the southwest, bringing in warm, dry air. The jet stream was producing jet streaks of faster velocity and causing excessive turbulence and uplift at the front. This unexpected situation developed rapidly and promoted tornado formation from multiple supercells.[31]

For this reason, on the morning of May 3, 1999, there was only a slight risk of severe thunderstorms forecast across the Central Plains. These forecasts followed computer models that only predicted a minor chance of severe weather because the jet stream influence was not included. However, by as early as 7:00 a.m., some meteorologists were becoming concerned about the developing conditions. Finally, they issued a forecast of potential severe weather at 11:15 a.m. and emphasized it in the 12:30-p.m. update. By 1:00 p.m., the development of tornadoes was declared to be a probability. Then at 3:49 p.m., the Storm Prediction Center in Norman, Oklahoma, forecast severe storms over the next several hours, and they issued tornado watches for most of Oklahoma, as well as parts of Texas and Kansas. A strong thunderstorm had already erupted by 3:20 p.m. from a huge supercell in Tillman County, Oklahoma. The storm crossed into Comanche County shortly after 4:00 p.m. and produced hail as large as 1.75 in. (4.4 cm). As a result, at 4:15 p.m., the National Weather Service issued a severe thunderstorm warning for the entire county, and it was upgraded to a tornado warning at 4:50 p.m. with the touchdown of the first of fourteen tornadoes that would form from this supercell.

The ninth tornado from this supercell formed at 6:23 p.m. in Grady County south of the town of Amber. It quickly strengthened into an F4 wedge tornado between one-quarter to one mile (0.40 and 1.61 km) across. It traveled for 6.5 miles (10.5 km) and struck the town of Bridge Creek, where it strengthened to F5. The tornado expanded to a width of about 1 mile (1.6 km) with the strengthening. It peeled 1 inch (2.5 cm) thick asphalt off roads, destroyed about 200 houses and mobile homes, reduced a well-built home with anchor bolts to bare slab, and killed twelve people in Bridge Creek. It was at this point that a mobile Doppler weather radar station measured winds of 301 miles per hour (484 kph) at a height of 105 feet (33 m) from this tornado, which is the highest wind speed ever documented.[32]

The tornado then tracked northeast past Newcastle in McLean County, where it destroyed thirty-eight homes and two businesses and damaged forty other homes. Several people sheltered under a highway overpass as had been made famous in a recent tornado video, but one woman was killed and others were injured because overpasses are actually bad places to shelter. The National Weather Service issued its first-ever tornado emergency for the Oklahoma City area at 6:57 p.m., and residents scrambled for shelter. The tornado entered South Oklahoma City at F2 intensity and a path width of 1/2 mile (0.8 km). It intensified to F4 and damaged fifty homes, sweeping several houses cleanly off their foundations. It continued to intensify to F5 and a path width of 3/4 mile (1.2 km), killing three people and causing extreme destruction.[33]

The tornado entered Moore, Oklahoma, where it reached F5 strength for the third time but weakened to F4 within the city. Some of the severest damage of the track took place in Moore, where several large housing complexes were destroyed, 11 people perished, and 293 were injured. It then reentered South Oklahoma City at F4 intensity, where it threw an 18-ton (16,000 kg) freight train car about 3/4 mile (1.2 km). The tornado passed into Del City and Tinker Air Force Base at F3–F4 intensity, killing six people and destroying hundreds of homes. It continued into Midwest City at F4 intensity, where it destroyed numerous businesses by hurling cars from a huge auto dealership into them. The tornado neared F5 intensity yet again as it entered a residential area and killed three more people. The Moore tornado finally dissipated in Midwest City, after one hour and twenty-five minutes of continuous destruction. It left a path 38 miles (27.4 km) long and up to 1.3 miles (2.1 km) in width.

The tornado was directly responsible for the deaths of thirty-six people, and five others died from indirect causes, yielding a death toll of forty-one. In addition, 583 people were injured directly by the tornado with 645 total injuries in the disaster. At least 8,130 houses, 1,040 apartments, and 260 businesses were damaged or destroyed by the tornado, and this produced 2.20 million cubic yards (1.70 million m^3) of debris. Damage costs were estimated at $1 billion, which was the first time a tornado exceeded $1 billion in damage. In 2024 dollars, the damage exceeded $1.8 billion.

The ferocity of the 1999 Moore tornado astounded even the most seasoned tornado experts, and action was quickly taken. Thanks to the advances made after the 1974 Super Outbreak, the warning systems operated flawlessly. It is estimated that residents had an average of thirty-minute warning before the tornado struck. Most of the warnings were from television in this event. Approximately 75 percent of the Moore residents sought shelter in safe locations as the storm approached.[34] The National Weather Service estimates that there would have been at least 600 deaths from the Moore tornado without the early warnings and safety precautions taken by the area residents.

The National Weather Service meteorologists issued a "tornado emergency" when they anticipated that the Moore tornado would strike a populated area. This forecasting level was rarely used prior to the 1999 Moore tornado. At this level, all residents are warned to seek safety in tornado shelters and safe rooms. Safe rooms are armored and anchored to withstand tornado winds up to 250 miles per hour (402 kph), and debris strikes at speeds up to 100 miles per hour (161 kph). However, in 1999, only 16 percent of impacted residents had access to a tornado shelter much less a safe room.[35]

In response to the 1999 Moore tornado, the Federal Emergency Management Agency (FEMA) teamed with the state of Oklahoma to offer an incentive program to encourage the installation of more safe rooms. FEMA initially provided up to 75 percent of the cost of safe rooms for qualified residents. Oklahoma and FEMA provided $57 million through this program, supporting the construction of at least 11,386 private safe rooms and 382 safe rooms in public buildings so far.[36] The protection levels were also increased in many existing shelters and safe rooms in response to the Moore tornado. The city of

Moore contains the highest per capita number of safe rooms in Tornado Alley and the United States as a result. In Moore, there were 3,236 registered safe room shelters by 2013.[37]

Also in response to the 1999 Moore tornado, FEMA assembled a Building Performance Assessment Team (BPAT) to assess damage in the tornado path and to make recommendations on building construction. This was the first time that a tornado disaster was assigned a BPAT investigation.[38] The city of Moore also examined building codes after this, and two succeeding deadly tornadoes caused significant damage to the city. On March 14, 2014, the city adopted enhanced building codes to prevent or lessen tornado damage. They raised the wind load standard to 135 miles per hour (217 kph) from 90 miles per hour (145 kph). These changes required significant changes to building houses in the city.

The Federal government also took action directly as the result of the 1999 Moore tornado. On January 14, 2000, the Housing and Urban Development (HUD) department began allowing borrowing for windstorm shelters. Further, the Tornado Shelters Act was passed and signed into law on December 3, 2003. This provides funding to improve protection against windstorms of all types on a national basis.[39]

7

Tsunamis

Tsunamis are deadly waves generated in the ocean by earthquakes, underwater landslides, surface to water slides, volcanic explosions, and extraterrestrial impacts. Earthquakes are by far the most common source. The cause of most tsunamis is a rapid change in the elevation of the seafloor that lifts or drops the column of seawater above it. Water cannot have a step on its surface, and it re-equilibrates the flat surface in a tsunami wave. Tsunamis caused by land-to-sea avalanches can be analogous to splashing water if they act on the water surface instead of the whole column. Volcanic explosions expel the water away in all directions through rapid expansion. Extraterrestrial impacts can have aspects of all mechanisms.

Once generated, tsunamis travel through deep water at phenomenal speeds of about 600 miles per hour (965 kph). This high speed allows them to cross entire ocean basins in several hours. In deep water, they have very low heights, usually well less than 3.3 feet (1 m), but they are very broad with widths of 300 feet (91 m) or more. The broadness of these waves means they displace an enormous amount of water compared to typical ocean waves regardless of their height.

In shallow water, tsunamis behave very differently. Water motion from waves penetrates the ocean to a depth of one-half the distance between waves. Water motion from a tsunami wave with a width of 300 feet (91 m) therefore penetrates to a depth of 150 feet (46 m) below the surface. The depth between water motion and no motion is the wave base. The seafloor is shallower closer to the shoreline. As the tsunami approaches the shore, its wave base drags on the seafloor, and the friction causes the wave to slow down. The front of the wave slows quicker than the back, which compresses the wave into a thinner but much taller wave. This is exacerbated if the coastline funnels the wave into a tighter area like a harbor. For this reason, the word "tsunami" means "harbor wave" in Japanese. The reason is that on a day with clear weather and calm seas, a monstrous wave will rise up out of the harbor without warning and with no apparent source and wreak havoc on a community.[1]

As a tsunami approaches the shore, the wave contains so much water that it pulls the seawater away from the shoreline as far as several hundred feet depending on its size. This is called drawback, and it exposes normally submerged seafloor. Unfortunately, this has the tendency to draw the attention of onlookers who inspect this oddity rather than running away. Drawback can precede the tsunami wave by up to three to four minutes. The tsunami then strikes at peak height and travels at 50–60 miles per hour (80–96 kph). The curiosity seekers cannot outrun this wave and are consequently become its first casualties.

Tsunamis can hit the shoreline with heights of up to several hundred feet locally. The record height was from an earthquake in 1958 in Lituya Bay, Alaska, that shook loose an avalanche that hit the water so hard that it drove a tsunami 1,750 feet (533 m) up the facing slope. Fortunately, no one lived there. More commonly, the tsunami strikes the shore at a height of 40 feet (12 m) or less. As it hits the shore, it quickly loses its height, but the water volume and momentum drive it well inland where it still causes devastation. The farthest point that the tsunami water penetrates inland is the tsunami reach or inundation limit, and it can be as much as 6 miles (10 km) if the topography is flat. Once the water is on land, it flows quickly back out to sea in the process called flowback. This flowback can drag debris and people many miles offshore, much stronger than a rip current.

A microtsunami is too small to see and can only be detected by sea level monitors. If a tsunami only strikes the coast right around its source, it is a local tsunami. If the tsunami is powerful enough to impact other areas around an ocean, it is a remote tsunami in those areas. Tsunamis that cross an entire ocean and impact opposite shores are teletsunamis. Very large tsunamis are referred to as megatsunamis.

365 CE Eastern Mediterranean Tsunami

Many tsunamis are produced by earthquakes. Determining which part of the disaster, the earthquake or the tsunami, is more impactful on history can be a matter of debate. However, regarding the 365 CE earthquake and tsunami, although the earthquake was exceptionally powerful, the tsunami was clearly the most impactful. In fact, this natural disaster was so impactful that it was commemorated as the "day of horror" for more than 200 years and remembered for more than 1,000 years. It was referenced by at least thirty authors in the fourth through fourteenth centuries.[2] This was a widespread event that occurred at an important time in history. There is debate as to how much it contributed to the division and ultimate downfall of the Roman Empire, but an accelerated period of decline followed it, followed by the splitting of the eastern empire. Further, it occurred at a time when paganism and Christianity were competing for public support, and it had an impact in that regard as well.

The epicenter of the 365 CE earthquake occurred near Crete as a result of convergence between Africa and Europe on the small Hellenic subduction zone arc in the eastern Mediterranean Sea area. Although this is a subduction zone which might be expected to produce a megathrust earthquake, the 365 CE earthquake was on a steeper fault within the overriding plate.[3] Tsunami modeling shows that ruptures on a normal fault, as proposed for this event, produce stronger tsunamis that better match the historical records for the 365 CE event.[4]

Early in the morning of July 21, 365, CE,[5] a very powerful earthquake with a magnitude of between 8.3 and 8.5 struck an area just off the west coast of Crete. The earthquake was likely the most powerful seismic event in the history of the Mediterranean region. It was produced by rupturing of a normal fault with a 30° slope angle. The ruptured segment had a length of some 60 miles (100 km), producing a slip of about 65 feet (20 m) to a depth of about 28 miles (45 km) beneath the surface. Surface features and reported damage in western Crete indicates that the earthquake had a minimum shaking intensity of XI on the Modified Mercalli Scale, which is extreme.[6] Historical reports about the earthquake indicate that it was felt everywhere throughout Eastern and Western Mediterranean and as far away as Spain. There was reported strong shaking in Dalmatia, Sicily, Libya, Cyprus, Egypt, and Palestine. It was likely to have been felt as far away as the Po plain in Italy and the city of Grenoble about 1,120 miles (1,800 km).[7]

The faulting caused great uplift and depression of the ground surface. Geological mapping shows that parts of western Crete were uplifted as much as 33 feet (10 m) above the sea level at the time by the faulting. The whole area showing uplift is about 100 miles (160 km) by 50 miles (80 km) with uplift ranging from 0 to 30 feet (9 m). However, some researchers have interpreted the uplift to have occurred as the result of two or more earthquakes over a period of two to three centuries rather than a single event.[8] This was a very seismically active period with strong earthquakes reported across southern Europe throughout the Middle East.

However, this record-breaking earthquake appears only to have produced excessive damage and fatalities in the epicentral area. The real history-changing aspect of the event was the tsunami it produced. The rapid uplift in Crete also uplifted the seafloor generating a major tsunami. This wave inundated coastal areas all across North Africa, the Adriatic Sea, Greece, Sicily, and the Middle East, especially on the Peloponnesus, the Greek Islands, Sicily, Libya, Cyprus, Palestine, the Nile Delta, Careen, Apollonia, and elsewhere along the

North African coast.[9] In the Nile delta area, the tsunami is reported to have devastated both the Alexandria harbor and city. Ships were reportedly tossed onto the roofs of houses and carried up to 1.9 miles (3 km) inland by the waves. Numerical modeling of the wave heights of the tsunami indicates that it was greater than 26 feet (8 m) in Alexandria.[10] Vast, rich agricultural lands of the Nile Delta were reportedly abandoned as a result of the flooding and saltwater incursion. Also, after the disaster, several previously populated cities in the Nile Delta were abandoned and subsequently only inhabited by hermits.

The Nile Delta was not the only area that showed the devastation of the tsunami. In southwestern Crete, at least two pulses of tsunami waves were identified in sediments in the coastal areas. A third wave pulse is also identified in some areas.[11] The inundation from the tsunami appears to have reached more than 525 feet (160 m) inland of the rocky coast and topped cliffs more than 33 feet (10 m) high. Sediment deposits at El Alamein and Kefr Saber are also from the 365 CE tsunami. Deposits all along the coast of Tunisia contain 16–28 inch (40–70 cm)-thick jumbled tsunami deposits[12] as do the coastlines of Libya and Egypt. It is estimated that the tsunami had a height of about 10 feet (3 m) in these areas. In the deeper sea, there is evidence that the 365 CE tsunami caused huge underwater landslides and debris flows, leaving deposits 50 feet (15 m) thick of jumbled sediment.[13]

The death toll from this massive event is roughly estimated at 50,000 people.[14] Considering the number of people that can be washed out to sea by a tsunami combined with poor record keeping and efforts in accounting for the dead, this number could be greatly in error. However, Roman historian Ammianus Marcellinus called it the "destruction of all the world," and other historians of the time labeled it a disaster of biblical proportions.[15] Many cities that were destroyed in the disaster were never rebuilt, and even cities that were less damaged did not recover. Excavations in Crete found that a large number of people died but were not buried. This indicates a large-scale breakdown of society. The city of Knossos was essentially leveled by the earthquake and tsunami, and the majority of the cities on Crete that were badly damaged were also abandoned.

The timing of the 365 CE disaster was a major factor in why it was so impactful. The Roman emperor Constantine converted to and accepted Christianity as the new religion of Rome in 313 CE. However, the transition was not smooth. Paganism and Christianity existed side by side for years but not very harmoniously. The emperor Julian ruled just prior to the 365 CE disaster from 361 to 363 CE. Although he was victorious in several battles against invading barbarians, he also attempted to restore paganism against the wishes of many Romans. As a result, the disaster was viewed by many as God's punishment or as God's sorrow for Julian.[16] In addition, the earthquake and tsunami destroyed many pagan temples and Christian churches as well as numerous other dwellings around the eastern Mediterranean. However, where rebuilding took place, pagan temples were not rebuilt, but churches were. This produced a striking change from pre-disaster to post-disaster in the appearance of religious affiliation. It now appeared after the tsunami, that the world was not mixed but firmly Christian.

The catastrophic impacts of the earthquake and tsunami are also believed to be significant catalysts in the decline of the Roman Empire.[17] Barbarian attacks on the northern lands of the Roman Empire frightened many of the citizens, and they fled their homes for the safety of Rome. The 365 CE disaster and related seismic events across the region caused the citizens there to also feel unsafe, and a mass exodus from the southern, Mediterranean part of the empire back to Rome also took place. The incoming population and their needs, the attacks by barbarians ever farther into the empire, and the fear of the citizenry combined to create great social instability. The social instability was coupled with political instability with frequent changing of emperors, every few months to a few years from about 355 to 380 CE. This overall instability led to civil wars, deep religious conflicts, and a widening gap between the eastern part of the empire headed in Constantinople and Rome. This situation finally culminated in 395 CE, just thirty years after the disaster, in the final division between the eastern and western Roman Empires.[18] The western, great Roman Empire would fall just eighty years later in 476 CE.

2004 Indian Ocean Tsunami

In addition to the power of the event, timing and location are the most important factors in a natural disaster changing history. Typically, this means time in history but in some cases, it can be as short term as the wrong day. This was the case with the 2004 Indian Ocean tsunami which struck Indonesia on December 26, 2004. The warm beaches around the Indian Ocean are winter vacation destinations for many tourists, primarily from Europe and Asia. The Christmas holiday week was the peak of the tourist season, leading vacationers to fill the resort towns to capacity all around the Indian Ocean. Christmas dinner and parties kept vacationers busy on Christmas Day, so they were eager to relax on the beaches the next morning. They were just leaving their hotel rooms and heading to the beach when at 7:58 a.m., a phenomenal submarine earthquake of magnitude 9.2 struck the seafloor about 100 mi (167 km) off the western coast of Banda Aceh in Northern Sumatra, Indonesia. The focus was 19-mi (30 km) deep. This was about as powerful as the second strongest earthquake ever recorded.[19]

The earthquake was generated by a megathrust fault of the Sunda trench subduction zone. This convergent margin is where the Indian Ocean and Asian plates collide. The fault slipped up to 98 feet (30 m) upward along a huge 720–808-mile (1200–1300 km) fault segment, the longest ever measured. Faulting progressed from south to north for a reported 8.3 minutes. This is the longest duration of faulting ever measured. Most of the movement occurred along a 240-mile (400-km) segment of the fault. The fault movement lifted the seafloor by the same amount as the slip, as well as all of the ocean water above it. The redistribution of water in the ocean to accommodate this uplift that formed a huge teletsunami that devastated the entire Indian Ocean basin.[20]

Seismologists at the US Geological Survey (USGS) monitoring station realized that this was an immense earthquake as the wave arrivals were recorded on their seismographs. Because the earthquake epicenter was out at sea, they knew that a major tsunami could be produced. The problem was that unlike the Pacific Ocean, which had the Pacific Tsunami Warning Center (PTWC), a real-time tsunami monitoring system that was installed after major tsunamis devastated the coasts in the early 1960s, the Indian Ocean did not. The PTWC not only senses tsunamis but also sends out a warning to all coastal communities who have been educated on what to do in the event of a tsunami. There was no warning system around the Indian Ocean at that time, so no alerts could be sent out, and the residents wouldn't know what to do even if they were warned. To warn the residents, the USGS would have to call every coastal town around the region, which was impossible. Even if it was possible to make that many phone calls such a short time, at 7:30–8:00 in the morning on Boxing Day, right after Christmas celebrations, very few people would be available to receive them. As a result, this massive tsunami struck these overbooked coastal resorts with no warning at all.[21]

The huge tsunami waves first struck nearby Northern Sumatra just ten to thirty minutes after the earthquake.[22] The maximum height of the waves is reported at 167 feet (51 m) locally but commonly at 66–98 feet (20–30 m). Residents report that about ten–twelve waves struck the shore. The first wave came from the southwest at about 1.6–8.2 feet (0.5–2.5 m) high but the second and third waves were 49–98 feet (15–30 m) high and devastating. This surge penetrated at least 0.6 miles (1 km) inland in some areas.[23]

In an hour and forty-five minutes to two hours, the tsunami reached the resort beaches of Thailand. There are many tourist videos of the ocean drawn back from the beach by as much as 100 yards (91 m) at places like Patong Beach for several minutes, and curious people wandering out on the exposed seafloor for a look. In the distance, a white line of furiously foaming water is hard to discern at first but quickly appears as a 33-foot (10 m) wall of water racing toward the shore. At that point, everyone on the beach began to panic and sprint away from the incoming terror. This was not the only area of Thailand to be impacted. The province of Phang Nga accounted for 71 percent of the deaths in Thailand. Khao Lak beach experienced 18–20-foot (5.5–6 m) waves along the western coast, and eyewitness accounts report that a 6.5-feet (2 m)-high surge flooded the area for well over an hour.[24] On Maikhao Beach in Phuket City, a ten-year-old British tourist recently studied tsunamis in school and recognized the receding ocean as a precursor to a tsunami. She warned as many people on the beach as she could to evacuate the beach, and as a result, she saved many people.[25]

About two hours after the earthquake, the island of Sri Lanka was struck by the tsunami. Even though it is 1,100 miles (1,700 km) from the epicenter, the waves were still up to 41 feet (12.5 m) high depending on location. Further, the inundation area reached as much as 1.2 miles (2 km) inland depending upon location.[26] The tsunami also reached the Indian mainland about two hours after the earthquake, and it wreaked havoc there as well. In one case, the tsunami is reported to have reached the third floor of a resort hotel. Among the worst cases in India, the city of Karaikal was ravaged by a 33-foot (10 m)-high black and muddy tsunami, and it killed 492 people. The tsunami next struck the Maldives 1,600 miles (2,500 km) away, beginning in a bit more than two hours. The highest tsunami waves there were about 13 feet (4 m) high.

This was a teletsunami, so it crossed several ocean basins with persistently impressive wave heights. In about eight hours after the earthquake, it struck Somalia in Africa more than 3,107 miles (5,000 km) away with wave heights of almost 33 feet (10 m).[27] About sixteen hours after, it struck South Africa with a 5-foot (1.5 m) surge. It then registered on Japanese tidal gauges in Antarctica with waves still up to 3.3 feet (1 m). Tsunami waves were even measured in North and South America at heights in the range of 7.9–15.7 inches (20–40 cm), but in Manzanillo, Mexico, the tsunami waves came ashore at about 8.5 feet (2.6 m) in height. It has been interpreted that the tsunami was channeled along deep ocean ridges, accounting for the vast variation.[28]

The death toll for this disaster was difficult to determine, even in modern times when records are so much better than ever. The problem was that the waves were so powerful that they didn't just kill individuals, they struck coastal communities and removed entire neighborhoods if not the whole communities. They abraded away so much sand and sediment that even the land beneath the community was removed, and the location of the old community wound up offshore. As a result, the death toll estimates varied wildly for years, reaching as high as more than 292,000 fatalities in twelve countries. The current estimate is that 227,899 people lost their lives in the tsunami, still making it the deadliest tsunami in history. By far, the greatest death toll was in Indonesia where 167,736 people were killed, followed by Sri Lanka where 35,322 people lost their lives. Thailand had a much lower death toll of 8,212, and the highest death toll in Africa was in Somalia where 289 people were killed.[29] Many of the deaths included tourists from several European and Asian countries making the countries impacted extensive. Even the survivors suffered greatly from the disaster. It is estimated that some 1.7 million people were left homeless, many of whom had lost family members and loved ones. It took years to get everyone settled into permanent accommodations. The total damage of this disaster is estimated at $13 billion.

Considering the advancements in smartphones just prior to this time, the Indian Ocean tsunami had the best media coverage of any other natural disaster to that point in time. Television and the internet had amazing tsunami pictures and videos for weeks to months after the event. Most were amateur videos recorded by tourists, and they frequently ended with a mad scramble by both the videographer and the people around them as the wave approached. Many of the photos showed absolute devastation of whole towns and coastal areas of cities. The bays, harbors, and lagoons had so much debris floating in them that a person could walk on it. Mixed with the debris were hundreds of human and animal corpses decaying in the hot sun. It was truly a gruesome sight.

The Indian Ocean tsunami may have been a surprise to the residents around the Indian Ocean, but it was not a surprise to others. The Pacific Tsunami Warning Center (PTWC) in Hawaii issued an alert for a possible tsunami just fifteen minutes after the earthquake. The reason for this is that an international tsunami warning system was installed in the Pacific Ocean soon after the massive 1960 Valdivia, Chile earthquake and tsunami. The 1960 earthquake was the strongest ever measured at a magnitude of 9.6, and the massive tsunami was just one of many that devastated the Pacific coastal communities. The Pacific Tsunami Warning System, of which the PTWC is part, was developed and installed by an International Coordination Group under the UNESCO Intergovernmental Oceanographic Commission.[30]

The PTWS uses Deep Ocean Assessment and Reporting of Tsunamis (DART) technology. DART is a system that warns of approaching tsunamis while they are still far from coasts. DART includes seafloor seismic sensors that detect tsunamis. These sensors detect changes in the weight of water above them, which increases as the massive waves pass over them. Once detected, an alert is sent to a buoy on the ocean surface. The buoy beams the alert to a Geostationary Operational Environmental Satellite (GEOS) above the region that broadcasts a warning to stations all around the Pacific Ocean. Coupled with a global seismic network to detect the tsunami-producing earthquakes, DART warns all coastal communities about incoming tsunamis early enough to allow evacuations. Evacuation routes are well-marked, and residents are educated on how and when to use them. The United States operates the Pacific Tsunami Warning Center in Hawaii, the Alaska Tsunami Warning Center, and the West Coast Tsunami Warning Center in California. Automatic earthquake alerts are sent to the centers for all earthquakes with magnitudes of 6.5 or greater, and warnings are issued for earthquakes with magnitudes of 7.5 or greater, except for Alaska where all magnitudes greater than 7.0 generate warnings.

The 2004 Indian Ocean tsunami and its staggering death toll were the impetus to initiate the Integrated Research for Disaster Reduction (IRDR) program under the International Council for Science (ICSU) in conjunction with International Social Science Council (ISSC) and the UN International Strategy for Disaster Reduction (UN-ISDR). Already by January 2005, the Hyogo Framework of Action (HFA) for 2005–2015 was developed at the World Disaster Reduction Conference and was endorsed by the UN General Assembly. This framework aims to proactively build the resilience of nations and communities to natural disasters of all types including tsunamis. As a result of these efforts, tsunami early warning systems have been implemented in the Indian Ocean, the northeastern Atlantic Ocean, the Mediterranean Sea, and the Caribbean Sea. In this way, the 2004 Indian Ocean tsunami has changed history by making many people and coastal communities worldwide much safer from deadly tsunamis in the future.

Natural disasters can also weigh heavily on existing conflicts, and they are more frequently followed by increased unrest. Further, even if resolutions are reached during the disaster, they rarely last. The 2004 Indian Ocean tsunami struck two areas of conflict and impacted both of them. Banda Aceh in northernmost Indonesia and Tamil Eelam in northeast Sri Lanka have been plagued by secessionist conflicts for the past three decades. The Acehnese people have been struggling for independence since 1873, when the Dutch took control of the area. After independence, the pro-Islamic Acehnese forces have been battling the secular, centralized Indonesian government, especially during the 1950s and 1960s. In 1976, the conflict intensified pitting the Free Aceh Movement or Gerakan Aceh Merdeka (GAM) against the Indonesian security forces. Intense fighting resulted in the deaths of more than 15,000 people and widespread accusations of human rights violations. In 1998, the Government of Indonesia passed autonomy legislation in an attempt to end the conflict, but it did not suit the GAM agenda. After two ceasefire agreements were broken by GAM during these negotiations, the Indonesian president declared Martial Law in Aceh in May 2003. The Indonesian security forces sent 40,000 troops and 12,000 national police to enforce the martial law and defeat the insurgents. The resulting hostilities displaced more than 125,000 people.[31]

The Indian Ocean tsunami devastated the Aceh area and a few nearby islands, but the rest of Indonesia escaped relatively unscathed. At this point, the Indonesian government could have mounted an offensive and effectively eliminated the now weakened GAM threat. Instead, on the following day, the president declared the tsunami damage to be a national disaster (not just Aceh) and called for the unification of the country to recover and heal from its previous conflicts. In response, GAM agreed to a unilateral ceasefire to facilitate the extensive relief operations in the Aceh Province. The security forces would aid in these operations but also ensure safety in the region from looting. Despite these overtures, sporadic hostilities and military operations primarily against GAM continued for a month after the tsunami when a comprehensive settlement was reached. The peace negotiations were facilitated by a former Finnish president to ensure neutrality. As a result, GAM agreed to drop its demand for independence if the region could have self-rule with monitoring by a committee from the European Union. After GAM and the government of Indonesia signed a Memorandum of Understanding on August 15, 2005, the House of Representatives enacted a new law on Aceh governance. In December 2006, Acehnese parties won provincial elections, and a former GAM member became the first democratically elected governor in the province. He pledged to address any remaining unresolved issues in the peace negotiations. However, for all intents and purposes, the natural disaster led to peace in a region of significant conflict.

In Sri Lanka, the Tamil secession conflict began in the 1970s. It began with negotiations to divide power between the majority Sinhalese and minority

Tamil which failed. This failure led to sectarian riots. The conflict deteriorated into civil war by 1983 between the Tamil secessionists and the Government of Sri Lanka, which was dominated by the Sinhalese. Muslim communities evicted from Tamil territories in 1990 also became caught up in the conflict. It is estimated that about 60,000 people were killed in the conflict before both sides agreed to a ceasefire in 2002. In addition, more than 375,000 people were displaced by the conflict. Since 2002, peace negotiations had been underway under the Sri Lanka Monitoring Mission (SLMM) led by a Norwegian contingent, but the peace process had stalled. This was the situation when the tsunami struck.

Unlike in Indonesia, the disaster impacted most of Sri Lanka and virtually all communities. The most devastated areas by the tsunami were in northeastern Sri Lanka where the prior hostilities were at their worst. There was an initial cooperation of the factions for relief, recovery, and rebuilding of the impacted areas. However, all parties soon began operating opportunistically to take advantage of the disaster to better their position. This was not just the Sinhalese and the Tamils but a Karuna faction in the east as well. As a result of the lack of good faith efforts to resolve the conflicts but instead to take advantage of the disaster, the ceasefire was broken and hostilities once again erupted in Sri Lanka.

8

Droughts and Famines

Virtually all terrestrial life depends on clean, fresh water. If the supply of water decreases, it stresses the natural ecosystems and human communities and can lead to significant loss of life in both. For this reason, droughts can be natural disasters. Droughts are periods of insufficient precipitation and streamflow in an area. There are systems or indices to classify the severity of droughts, and all compare the current amount of available water in an area to normal conditions.[1] Droughts are monitored worldwide to varying degrees depending on the country. The United States maintains an online real-time drought monitoring system.

The Palmer Drought Index was developed in 1965 and measures drought and wet periods that reflect the balance of water supply and demand. The Palmer Index was improved to the Standardized Precipitation Index (SPI) which measures precipitation as well as the balance between water supply and demand. The SPI covers several time periods from one to twenty-four months and assesses short-term and long-term droughts. The United States Drought Monitor system combines the SPI, the Palmer Drought Index, and US Geological Survey stream flow data to report drought impacts in near real time. It assigns drought indices of D0–D4 to all areas of the United States. D0 is abnormally dry relative to historical averages and includes short-term dryness and water deficits. D1 is moderate drought and includes crop damage, water shortages, and voluntary water restrictions. D2 is severe drought including crop loss and local water restrictions. D3 is extreme drought including major crop losses and widespread water shortages and restrictions. D4 is an exceptional drought and includes widespread crop losses and water emergencies.[2]

The US Climate Extremes Index (CEI) was developed in 1995 to quantify the climate changes in the United States. The CEI evaluates a set of climate measures including extremes in average monthly temperatures, heavy precipitation events, drought severity, number of days of precipitation, and wind speed of hurricanes and tropical storms at landfall. A CEI of 0 percent means that none of the United States experienced the extreme temperatures or precipitation. In contrast, a CEI of 100 percent means that the entire country experienced extreme conditions for every indicator. Typically, CEIs more than 20 percent indicate "more extreme" or above-average conditions, and CEIs less than 20 percent indicate "less extreme" conditions. The CEI is calculated for eight periods including annual, spring, summer, fall, winter, cold season, warm season, and hurricane season.[3] The CEI indicates that extreme conditions in the United States have been increasing since the early 1970s.

Drought is mainly caused by the lack of rainfall, and as a result, vegetation slowly dies. This causes several negative effects. The dry vegetation is much more flammable than moist vegetation, and wildfires are more common. The removal of vegetation leaves less to be eaten by animals, and they starve and die as a result. This lack of food impacts humans as well, and droughts are commonly accompanied by famines which can kill people. Vegetation also holds soil together. Its removal and the drying of the soil allow the wind to remove it from the surface. The danger is that the top layer of soil typically contains nutrients that make the soil productive. In windstorms, removed surface material can create dust storms. Dust storms can be hundreds of miles across, cause extensive damage, and pose public health dangers.[4]

Megadroughts are severe droughts that last for twenty years or more. There are several megadroughts around the world at all times, and this situation has persisted for millennia. A 1,200-year history of megadroughts has been determined for the American Midwest and West using tree-ring studies. There were four major megadroughts during this time. The 940–985 CE drought lasted 45 years, the 1100–1247 drought lasted 147 years, the 1340–1400 drought lasted 61 years, and the 1550–1600 drought lasted 50 years. The tree-ring study also showed that the twentieth century was the wettest century in the entire 1,200-year record.[5] Therefore, the wet weather that facilitated the settlement and development of the western/southwestern United States was a historical fluke. Consequently, it is probable that communities in these areas are not sustainable. This is not the only drought. The Australian drought that ended in 2019 was the worst in 800 years. The current drought in Brazil is the worst in history. Climate change is making all droughts worse.

Drought and Downfall of the Mayan Empire

It would be a reasonable assumption that drought disasters occur in areas that tend to be dry in general but that isn't always true. The great Mayan empire was in the current Guatemala, Belize, Yucatan Peninsula of Mexico, and parts of Honduras and El Salvador. This area is primarily tropical rainforest with current annual rainfall totals up to 175 inches (444.5 cm) per year with as much as 40 inches (102 cm) of rain in a single hurricane. The idea that such wet area could suffer from a drought disaster is unimaginable. However, it was drought that was a major contributor to the fall of this sophisticated civilization that had been in existence for millennia.

The neolithic Mesoamerican Mayan civilization may have been the most sophisticated in the ancient Americas. It was never really a single cooperative like a country, but instead it was a group of city-states and rural areas that developed local alliances within the region, similar to ancient Greece, throughout its history as far as is known. The Mayan civilization is historically divided into the archaic, Preclassic, Classic, and Postclassic periods, and contact with Spanish explorers began in 1511.[6]

<table>
<tr><th>PERIOD</th><th colspan="2">SUBDIVISIONS</th><th>TIMES</th><th>DROUGHTS</th></tr>
<tr><td colspan="3">ARCHAIC</td><td>8000–2000 BCE</td><td rowspan="5">?</td></tr>
<tr><td rowspan="6">PRECLASSIC</td><td colspan="2">EARLY</td><td>2000–1000 BCE</td></tr>
<tr><td rowspan="2">MIDDLE</td><td>EARLY</td><td>1000–600 BCE</td></tr>
<tr><td>LATE</td><td>600–350 BCE</td></tr>
<tr><td rowspan="3">LATE</td><td>EARLY</td><td>350–1 BCE</td></tr>
<tr><td>LATE</td><td>1 BCE–159 CE</td><td rowspan="2">~120–200 CE</td></tr>
<tr><td>TERMINAL</td><td>159–250 CE</td></tr>
<tr><td rowspan="3">CLASSIC</td><td colspan="2">EARLY</td><td>250–550 CE</td><td></td></tr>
<tr><td colspan="2">LATE</td><td>550–830 CE</td><td rowspan="3">~700–1125 CE</td></tr>
<tr><td colspan="2">TERMINAL</td><td>830–950 CE</td></tr>
<tr><td rowspan="2">POSTCLASSIC</td><td colspan="2">EARLY</td><td>950–1200 CE</td></tr>
<tr><td colspan="2">LATE</td><td>1200–1539 CE</td><td>1450–1539 CE</td></tr>
</table>

Spectacular Mayan monuments and even pyramids are tourist attractions in the forests of the Yucatan Peninsula of Mexico. But the Mayan people and civilization were mysteries for many years shrouded in the tropical rainforests of Mesoamerica. However, over many years, archaeologists and anthropologists slowly unearthed evidence of this rich and sophisticated culture that dominated the area, deciphered their hieroglyphics[7] and later even learned about the life of common people using LiDAR imagery surveying and analysis.[8] Now, the history of this fantastic civilization is much better understood and appreciated.

The Mayans had advanced agricultural practices including terracing of slopes, canalized field systems, and dams and sluices to impound and release water as needed.[9,10] They deforested large areas of the jungle and planted crops such as maize (corn), beans, squashes, chili peppers, manioc, sunflowers, cotton, and others.[11] Some crops and even agricultural gardens were planted within urban areas with lower population density, fringing urban areas in peri-urban or suburban settings, or in rural agricultural areas located away from the urban centers depending upon the area and the time in the civilization. During times of extensive military activity, agriculture moved into or tightly around urban areas for safety, but otherwise there were no rules with each urban center maintaining its own arrangement. Evidence of agriculture in Mesoamerica shows that it began at least as early as 2500 BCE but flourished in the Classic period.[12]

Mayan cities first developed about 750 BCE but grew quickly. By 500 BCE, the cities had large centers with monumental structures like large temples with elaborate stucco façades, pyramid temples, ceremonial ballcourts, and structures for astronomical observation. These cities expanded and grew more sophisticated, peaking in the Classic and Postclassic periods. Mayan culture included sophisticated art in wood, ceramics, sculpted stone monuments, stucco, painted murals, and smaller jade and obsidian objects. The Mayans developed hieroglyphic writing by the third century BCE, and it appeared on buildings and in documents. It is considered the most advanced writing system of the pre-European Americas. These hieroglyphs recorded historical events and ritual knowledge. The Mayans began with a single root language that first appeared around 1000 BCE in the northwest Yucatán. That single language stem later diverged into more than thirty distinct languages during the Mayan civilization, and many have survived into modern times The Mayans developed a sophisticated series of interlocking ritual calendars to mark timing of ceremonies and to record historical events. Their astronomical calendars were so accurate that they measured the 584-day revolution of Venus to within just two hours. They also developed advanced mathematics that included perhaps the earliest instance of the concept of zero in human history. Large, sculpted monuments were constructed in the Classic period, and many city-states were linked by interconnecting transportation and trade networks by that time.

The Mayans interacted with neighboring communities and groups including the Olmecs, the Teotihuacan, and the Aztecs, and they influenced each other to varying degrees. The major cities grew to populations of 50,000–120,000 people, but they had large suburban and even rural areas linked to them, and they were further linked to networks of subsidiary sites so city-states included many more people. By the Classic period, these city complexes could have populations exceeding one half million people. The most powerful city-states in the lowland area were rival centers of Tikal and Calakmul. However, by the Late Preclassic period, there were several large cities in the Petén Basin, and all competed with each other for resources and power. At its peak, the Mayan civilization had more than 100 major city-states across the empire with as many as 19 million residents in total.[13]

The governmental structure of the Mayan city-states was based around a single ruler or ruling family with divine authority and a royal court to support them. This supporting bureaucracy could be quite extensive. The citizens were loyal to the rulers and paid taxes, performed labor, and deferred to them, primarily without complaint. They maintained a ceremonial culture steeped in tradition that occasionally included human sacrifice depending upon the importance of the ceremony. They were also generally a war-like people. There are carvings and hieroglyphs of great battles, and there is evidence of area populations contracting sharply and relocating to the urban center for a period of time during conflicts and then expanding during periods of peace.[14] The lack of a centralized government, alliances that constantly expanded and waned or were terminated, the war-like nature of the culture, and the bickering and arbitrary and capricious nature of the leaders did not promote the sustainability of the civilization. Further, there is evidence that the forest cover was effectively stripped for agriculture during the Classic and Postclassic periods.[15] All it would take is exceptional stressors like natural disasters to cause major disruptions in society. Local natural disasters such as hurricanes, volcanic eruptions, and earthquakes undoubtedly struck parts of the area as they still do today. But regional stressors like major droughts were more than the fragile culture could handle.

By the time of contact with the Spanish, the great Mayan civilization had devolved into several scattered warring tribes. It was clear even then that drought had plagued the empire. There were journals, oral recollections, and even a few survivors to describe the conditions of the last days of the great civilization. Extremely cold weather conditions began by about 1447, but the resulting devastating famine peaked from 1450 to 1454. Records claim that snow fell deep enough all across the Mayan region in 1450 to cause the collapse of buildings. Many people died of both cold and thirst, and vegetation barely grew. In 1451, there are reports of animals entering villages, killing people, and eating them. Even young adults were emaciated and died of starvation. In 1452, buzzards were reported to be ever-present across the countryside. Survivors sold themselves into slavery, and others just dug a hole, climbed in, and died. Lizards then devoured the corpses. By 1454, there were reports of survivors selling children for any food at all in addition to the other atrocities. Survivors contracted diseases, and most people believed that no one would survive.[16]

The last great Mayan city was Mayapan, the ancient capital city on the Yucatán Peninsula of Mexico and perhaps the capital of the entire Postclassic empire. Survivors had flocked to the city for protection and resources during the thirteenth through the mid-fifteenth centuries, building the population to about 20,000 people. However, even this great city fell to an attack in about 1458, and the residents were massacred. The city was abandoned after this. This disaster was imposed on the citizens after also suffering through the severe drought for the past few years, but extremely dry conditions for centuries. There is even evidence that this great city was invaded about 100 years earlier and suffered a major massacre. These attacks appear to have been instigated by the impact of the drought and famine on their rivals.

This documented final civilization-terminating drought, and famine was severe. However, it is not that surprising because this was the beginning of the Little Ice Age, and severely cold and dry conditions were impacting many places on earth. The great Mayan civilization, however, did not disappear because of a single drought. Recent studies have shown that the Mayans suffered through several droughts and even megadroughts throughout their history. In fact, the entire history of expansion, advancement, and contraction can be explained largely by a series of droughts throughout the civilization.

Ice core data show that there was a lowered solar radiation output for at least the 760–930 CE part of the history, but it may have been weaker earlier as well. Sophisticated climate modeling indicates that reduced solar radiation results in the failure of summer rains developing in Central America, thereby leading to drought. Several proxies have been investigated that demonstrate the droughts and extreme climate situations and their timing including sea-surface salinity fluctuations as determined from piston coring of sediments from the Caribbean Sea, sulfur and salt content in cores of lakebed sediments, and chemical studies of stalactites and stalagmites from a cave in the region. The marine sediments from the nearby Caribbean indicate that surface salinity was markedly lower during the periods 200 CE, 550–600 CE, 850 CE, 1000 CE, 1400 CE, and 1600 CE with a fifty-year margin of error in the timing. These ages correspond directly to periods of wars, abandonment of cities and regions, and deflation of the population especially in the Mayan Lowlands. The Preclassic abandonment occurred at 150–200 CE, the early to middle Classic Hiatus or transition occurred at 536–595 CE, the Classic period collapse occurred from 760 to 930 CE, and the Postclassic abandonment and downfall of the civilization as described was from 1450 to 1454.

The data from the stalagmites support the interpretations from the marine sediments. They indicate a continuously dry period from 700 to 1135 CE with peak droughts occurring at 754–798 CE, 871 CE, and 893–922 CE, with each drought increasing in severity. In lakebed cores, sulfur contents were generally higher from about 500 CE to 1000 CE, indicating a greater frequency of droughts. In the Preclassic period, there are increases in sulfur at about 480 BCE, 370 BCE, and 265 BCE. There are two sulfur peaks between 120 and 190 CE about the time of the Preclassic abandonment. The highest sulfur contents occur in the Terminal Classic with three peaks between 750 and 870 BCE. In the Early Postclassic, there are two peaks at 1000 and 1060. Based on all of these different sources of data, it is estimated that this apocalyptic drought was the worst in 7,000 years.

During the Preclassic abandonment, major cities all across the empire were greatly reduced or completely abandoned. The major city of the empire at that time was El Mirador, which was abandoned and not repopulated at all until the Late Classic period. The population in the Three Rivers region to the east decreased by at least 70 percent. After this drought, populations recovered, and most cities were rebuilt over the following 300 years. During the Hiatus, rebellion, war, and social unrest halted construction and art in the Mayan Lowlands, leaving a gap in the record. The major city of Río Azul was abandoned, and the population in the area decreased by 70 percent. The next 200 years experienced normal rainfall, and the population and culture peaked in the Mayan Lowlands. However, this booming period was followed by among the most devastating droughts in world history, destroying the great Classic Mayan civilization in the Yucatan Peninsula and the Southern Lowlands. Many areas were completely abandoned. Although the first stage of this collapse began around 750 CE, it subsided, and a second phase struck the west and southwest particularly hard at about 810 CE. Then around 860 CE, a third phase devastated the southeastern part of the empire. However, other intense periods of drought struck in 890 CE and in 909–910 CE. By the end of this apocalypse, only small Postclassic enclaves around stable sources of fresh water in the north, east, and central Peten survived and persisted. These small groups slowly rebuilt the Mayan civilization. By 1200 CE, populated areas had expanded enough to reinvigorate the civilization. This is when the northern state with a capital at Mayapan was formed.

Recent sophisticated archeological and LiDar studies show that different urban centers were more impacted by the droughts to different degrees and at different times. Populated areas contracted and expanded with each of the droughts, and some areas resisted the droughts much better than others.[17] The droughts further encouraged civil and regional unrest, which led to wars and overthrow of city-states. This means that there were periods of relative peace and stability during the times of normal precipitation, but during droughts, rulers were overthrown and replaced several times, and cities shifted dominance and populations equally often.[18] The droughts impacted the Mayan civilization much more with the arrangement of small competing city-states than it would have if they had been working together as a single empire. Otherwise, resources could have been redistributed to areas of need. Therefore, part of the impact of the apocalypse was because of the governmental structure.

As is usually the case, there were many practices that exacerbated the impact of the droughts. Recent studies have found that local plants in Central America have a reasonable amount of drought resistance. However, agricultural crops have very little resistance. For this reason, when droughts struck, they were far more devastating than they might have been with native vegetation.[19] Another contributing factor was the massive deforestation that coincided with the population boom during the Classic period. The thick tropical rainforests retain water from precipitation for long periods, but agricultural fields do not. Therefore, the drought was greatly exacerbated for this reason as well.[20] The massive population boom during the Classic period also stretched the resources of the civilization far thinner than could be accommodated. All of these factors made a terrible situation even worse.

Drought and the Fall of the Great Ming Dynasty

Droughts commonly lead to famines which can be deadly. In current times with global trade and the ability of countries to take on debt, although droughts are still common, famines are less common in developed countries. However, in the past, some of the deadliest natural disasters were the results of droughts and consequent famines. In many cases, the failings of a government exacerbated the effects of a drought and famine by not dealing with it directly or actually making it worse. As is the case with most natural disasters that changed history, timing, and the surrounding situation play a major role in the impact. For example, central and northern China experienced a devastating and deadly drought from 1876 to 1878 that took the lives of between 9 and 11 million people, but the government remained the same, and there was no lasting change in history.[21] This was not the case with the Ming dynasty, which fell in 1644 primarily as the result of a terrible drought and famine and the negative results they produced.

The earth entered a cooler period about 2.6 million years ago that has been characterized by regular ice ages. These ice ages include the expansion of the polar ice sheets far into the normally temperate latitudes and a severe lowering of sea level because so much water is tied up in the glaciers. The last ice age extended from about 22,000 years ago to about 12,000 years ago. During this event, ice sheets extended all the way from the North Pole to New York City in North America. These ice ages caused profound changes in the weather patterns and climate both globally and locally. We are now actually considered to be in an interglacial period, signifying that the next ice age could begin at any time, although the climate crisis makes that less likely. However, the earth has started on the path toward an ice age several times but just not completed the transition. One of these is called the Little Ice Age, which lasted from about the fifteenth century to the middle to late nineteenth century, depending upon location, and it had a profound impact on the global and local climates.

The Ming dynasty officially began in 1368 with the fall of the Yuan dynasty. That dynasty was controlled by the Mongols and was started by the great leader, Kublai Khan. The Ming (meaning brilliance) Dynasty began with instability and uneasiness but soon progressed into a period of dramatic transformation in the Chinese political and social order and became among the most remarkable and progressive periods in the history of China. The border with the Mongols, marked by expansion and reinforcement of the Great Wall, was stabilized for the next 200 years.[22] The empire stretched from Beijing in the north 1,200 miles (1,931 km) to Canton in the South and laterally from Shanghai in the east 1,000 miles (1,609 km) to Chengdu in the west. The population of China more than doubled during the Ming Dynasty to between 160 and 200 million people at its peak.[23]

The Ming society included the development of a powerful economy. It contained extensive market networks, a powerful agricultural sector, and impressive artistic industries. Agricultural advances included the construction of 40,987 ponds and dikes for irrigation to improve agricultural output. Farmers introduced several equipment innovations including water-powered plows and new agricultural methods including crop rotation, among others. International trade brought in new crops like corn, potatoes, and sweet potatoes, among others. Ming silk, porcelain (like vases), and tea were found in major cities worldwide at the time. The increasingly modern and monetized Ming economy was fed by an influx of silver from Japan, Europe, and the Americas. There was an influx of traders from Portugal and other countries during this time, as well as Jesuit missionaries. All of the commerce spurred the growth of an affluent gentry class, an intellectual movement, and advanced literature and plays. It was truly a golden age. The capital was moved from Nanjing to Beijing where the monumental "Forbidden City" was constructed. This complex served as the imperial palace continuously until 1912 when the Qing Dynasty fell. More than 10,000 households were relocated to build the population of the city of Beijing.[24]

An empire as impressive as this would be expected to last for many centuries. However, this period experienced the transition into the Little Ice Age, which caused catastrophic climatic changes, especially in North China. It appears that the average temperature in the area increased until about 1500 and then decreased sharply in steps until about 1655 with few increases. The temperature would not completely recover from this downturn until the late nineteenth century. As a result, the average temperature anomaly from the 1560s to the 1640s was −0.76 °F (−0.42 °C), and the average temperature fell by a devastating 2.12 °F (1.18 °C) from the 1610s to the 1650s. This dramatic cooling was accompanied by other severe climatic changes. It generated the Late Ming Weak Monsoon Periods, which were characterized by the weakest east Asian summer monsoons in 500 years in China. In fact, these periods are among the severest climate events of the past two millennia.[25] The reduction in summer rainfall during these events was at least 50 percent and commonly closer to 100 percent from 1638 to 1641. The periods appear to correlate directly with multiple interpreted El Niño events between 1550 and 1640.[26] It has been calculated that 1640 was the driest year in the region for the next 500 years, and 1641 was the second driest year in 200 years. In western Shandong in 1641, the Grand Canal dried up. However, the entire period from 1637 to 1643 is considered the worst seven drought years of the Ming dynasty and was named the Chongzhen Slough.[27] These near-continuous dry periods throughout the early seventeenth century comprise the Chongzhen Megadrought, which dominated the end of the Ming dynasty. This climate event was the severest drought in the previous 500 years in North China and possibly 1500 years.[28] A drought that lasts two decades or longer is designated as a megadrought.

However, this cooling and drought was not regular or uniform across China. There was a clear north-to-south, time-transgressive drought pattern beginning in northern China and progressing to Jiang-Huai in the south over an approximate ten-year period. However, it was not even just a purely drought event. The precipitation pattern during the weak East Asian summer monsoon event included flooding in the south but drought in the north, which has been seen in recent similar patterns. Intervening La Niña events are speculated to have increased the typhoons hitting South China, causing rainfall and tidal floods in the southern areas of Guangdong, Fujian, and Zhejiang provinces. This pattern is proposed to have produced the "Wanli Jiangnan Flood" from late 1608 to 1610, which was the most devastating flood disaster in 200 years.[29]

These climate changes wreaked havoc on the Ming empire. The drought severely reduced grain output in the empire. The grain yield per unit area in 1630 in most areas in China was estimated to be 20 to 50 percent of the output in 1580, and the average for the whole country was as low as 42 percent of the 1580 output. The output had already decreased from its peak by 1580. These reduced outputs caused severe food shortages, which resulted in sharp grain price increases in the early seventeenth century. The drought decimated the agricultural output of the northern military farms, which provided food to the northern frontier troops. Cooling further curtailed the growth of the grasslands, and the shortage of foraging caused the death of nearly 90 percent of the domestic animals that were not already slaughtered for food. During these famines, people ate the bark off of trees and grass seeds or any plants they could find, edible or not.[30] It is reported that many people resorted to eating soil, which is possible but causes so many negative health effects that they would only live for ten to fifteen days.[31]

The famine was so great that large numbers of people resorted to cannibalism. Of those reported cases in North China, cannibalism occurred an average of 6.62 times per year in every extreme drought year or the next year including 1560, 1586, 1587, 1615, and 1638–41. In years when there was severe famine, cannibalism was reported more than ten times per year.[32] In fact, it is estimated that cannibalism was far more common than just that which was reported. There are reports that when local magistrates apprehended outlaws, rather than jailing or hanging them, they had the criminals beaten to death and then allowed the local residents to eat them. It is reported that hundreds of criminals met their demise in this way. There are reports of parents selling their children for enough money to buy a few portions of rice or grain. Parents are reported to have abandoned their families to certain death just for the chance to survive. In other cases, both individuals and whole families are reported to have committed suicide rather than die of starvation. Other reports claim that roving gangs hunted down and killed and ate anyone they could find, that mothers killed and ate their own children and that older siblings killed and ate younger siblings,[33] although these are not verified and seem less believable.

As if the drought in the north and expanding south wasn't enough, the flooding in the south and between the drought years also wreaked havoc on the Ming population. During the late spring and early summer, heavy and sustained rainfall fell in the Yangtze River in the south-central Anhui province, followed in June and July by floods in the Zhejiang province in southeast in wet years. In Henan, there was flooding during the summer of 1642, and a catastrophic flood of the Yellow River in 1642 that destroyed the city of Kaifeng. This former capital of dynastic China was entombed in sediment, killing a reported 300,000 people.[34] All of the cool and wet weather encouraged the incubation and spread of diseases. Food shortages and high food prices made it difficult for a normal lifestyle in many parts of China. A large number of farmers and others went bankrupt, lost their homes, and became refugees moving around the empire looking for a stable situation. This helped spread the disease. Rats and other rodents were also starving and moved around the empire. One of the more virulent and widespread of these diseases was a plague of the Black Death, which ravaged the northern part of the Ming Empire, killing as much as 40 percent of the population.[35] The plague peaked from 1633 to 1644, beginning in Shanxi Province but spreading to most of the country. The early seventeenth-century extreme weather events led to the worst outbreak of the Black Death called the Chongzhen pandemic, which lasted from 1640 to 1644. However, it wasn't just the Black Death that afflicted people. Several epidemic diseases raged throughout the year including typhus, typhoid fever, malaria, dysentery, and smallpox. There were also unidentified diseases like the "wool plague," which could have been a fungus disease, or the "pimple plague" and "vomit-blood plague" that struck in 1643 and were likely the bubonic plague and pneumonic plagues, respectively. The plagues peaked in 1643, killing more than 200,000 people or 20 to 25 percent of the population in Beijing alone. Despite great medical practices and advances at the time, this period marked the second most deadly time of plague in all of Chinese history.[36]

The climate and disease stressors were probably enough to cause the downfall of all but the strongest of empires. However, the Ming government, bureaucracy, and attitude had already begun to decay, making the conditions for downfall ideal. The possible beginning of the change in attitude and intention was the institution of the Single Whip Law, during the middle of the Ming dynasty. This law changed the unit of tax collection from grain to silver, which furthered the monetary system toward silver. This law drove a temporary increase in trade with Europe, but it also contributed to the weakening of the Ming economy in the long run. Overall, it destabilized the tax system, diminished the economic power of the government, and allowed private citizens to become more powerful.[37] These changes caused the Ming society to become more decadent. Observers of the late Ming era believed society to be too bounteous for its own good. The merchants who had become rich with new money infiltrated the top levels of society formerly reserved for Confucian scholars. In fact, these new rich people passed themselves off as Confucian scholars with their money. This decadence wormed its way into the government as well, which weakened it by causing it to cater to rich people. The Single Whip Law led to a decline in the granary system, which led to a decline in grain stores. As a result of the decline, grain was no longer widely used in disaster relief, which increased famine in the less affluent. Further, tax relief had previously been granted to hardship cases due to famine, drought, or flood but was abandoned with the new policies. The Ming court received 239 requests for tax relief between 1573 and 1590 and approved 115, or just under 50 percent. In contrast, it received 200 requests between 1600 and 1620 but only approved 40, or 20 percent. By the seventeenth century, governmental disaster relief efforts were essentially negligible. In response to severe droughts from 1627 to 1643 and the lack of government assistance or even much interest, the Ming peasants launched an uprising that would help destroy the empire.[38]

The peasant uprising was not the only pressure that the empire faced. Pirates primarily from Japan attacked ships carrying international traders, which reduced trade and income. They also attacked coastal communities and launched marauding forays far inland to plunder towns and cities. The Mongol threat along the northern border never abated throughout the entire dynasty. A new threat emerged from the northeast as the Manchu tribes who were also suffering from the effects of the cooling climate, and drought joined forces to form powerful new military groups. They too threatened the empire in the late sixteenth and early seventeenth centuries. The drought and external threats increased the military budget from 64 percent of total Ming budget in 1548–69 to 76 percent in 1570–89. The severe droughts from 1627 to 1643 then destroyed the military farm system, which supplied provisions for Ming troops along the northern frontiers. Under the new pressure from the peasant uprising and Manchu invasion, the government increased taxes in 1618–20, 1637, and 1639 to pay for the military costs, which increased taxes several fold. However, the diversion of funds from the Ming empire could not keep up with the needs of the military. The Manchu destroyed the main part of the Ming army in several decisive battles.[39] The rebel forces from the northwest and suffering from famine numbered more than 30,000 and were led by the Robin Hood-like "Dashing King." They attacked the remaining 50,000 soldiers defending the capital city of Beijing and captured it. The last Ming emperor committed suicide rather than be captured. After their victories and with no substantial opposition, the Manchus crossed the Great Wall, en masse, and absorbed the entire country of China into their new Qing dynasty, which began with the fall of the Ming dynasty in 1644. It is estimated that at least 26 percent of the Ming population perished in the struggle, famine, and plague in the downfall. At least 10 million people perished from the drought and famine and up to another 10 million perished from the plagues. Only 32 percent of the 1630 population survived in North China.

Recent studies have focused on six factors that are identified as leading to the collapse of the Ming dynasty including external wars, internal rebellions, interethnic conflicts, drought, floods, and locusts. The first three are human factors and account for about 47 percent of the downfall. However, the last three are natural disasters and account for about 53 percent.[40] As with all natural disasters that change history, proper timing is required. If the drought had occurred when the Ming dynasty was still on the granary system and before it had become decadent and self-absorbed, with the government facilitating the growth of individual wealth instead of the well-being of the empire, the dynasty may have withstood the drought and famine. However, the finances were spread so thin and government corruption so widespread that there was no way for even the powerful Ming dynasty to withstand this natural disaster.

Drought and Dust Bowl

One of the best examples of human activity exacerbating a natural disaster is the Dust Bowl drought and disaster of the Midwest United States in the 1930s. This disaster destroyed the quality of the soil over hundreds of square miles of very productive land. It caused the largest short-term migration of people in the history of the United States. It also compelled the most sweeping legislation on agriculture in the United States. However, it was poor land planning in terms of lot size, poor farming techniques, and poor oversight of the land use on a local basis that greatly intensified the impact of the disaster on the environment and the people. As a result, the Dust Bowl is regarded by many as the worst natural disaster in US history.[41]

The Great Plains of the United States is a dry area bordering on arid depending on the climate at the time. The bedrock is covered by thick sediments made of alluvial deposits, fluvial deposits overlain by windblown loess, and glacial deposits to the north, all covered by soils made from the decomposition of these largely inorganic deposits. The deposits and soils have been covered by fields of tall grasses for tens to hundreds of thousands of years, depending upon location. These deep-rooted, natural grasses fix the soil and have evolved to survive the dry, cold, and windy conditions typical of this region. The dead grasses slowly decayed, producing the thin topsoil that mantled the sediments. The topsoil provided most of the organic carbon available for the vegetation across the area. The slow organic production and lack of appreciable precipitation make the area very sensitive to climate changes, as well as disruption of the natural environment.

The Great Plains were covered in these tall grasses and occupied by tribes of nomadic Native Americans through the first half of the nineteenth century. The big change came with the passage of the Homestead Act in 1862, which encouraged a mass migration to the area. It provided each settler or family with 160 acres (65 hA) of public land to farm. This was later expanded by the passage of the Kinkaid Act in 1904 and the Enlarged Homestead Act in 1909, which increased the size of the claims. As a result of these acts, a large number of new and inexperienced farmers invaded the Great Plains. These farmers plowed under the prairie grasses and turned the soil, leaving the topsoil exposed to the sun and wind. They primarily planted wheat but also corn, soybeans, and alfalfa. Some settlers chose to graze cattle on their land, which was less damaging than plowing but not by much because overgrazing also destroyed large areas of grassland. Demand for agricultural production grew with the First World War, compelling farmers to cultivate more land to produce more food.[42]

At the end of the Little Ice Age, the climate of the Great Plains entered an especially wet and warm period. For about a century, the area experienced above-average rainfall and moderating temperatures on average. This does not mean that there were no droughts. In fact, the drought of 1856–65 may have resulted in a sharper drop in precipitation than the Dust Bowl drought.[43] There were also periods of low rainfall in the area in 1890 and 1910. Recent studies suggest that the Dust Bowl was just one of a series of droughts occurring approximately every twenty years.[44] But the farmers were unaware of this possibility and were encouraged to ignore it by the recent precipitation trends. From 1918 through 1927, the annual precipitation averaged more than 24 inches (61.5 cm) in the western or high Great Plains in comparison to the usual 20 inches (51.5 cm).[45]

The response to mechanical advances in farming, wet weather, and the increased demand for wheat was a near doubling of the cultivated land from 1910 to 1920 in the Great Plains. More than 61,775 square miles (16 million ha) were cultivated for wheat production. All of this land was now exposed to potential erosion. Even less desirable areas like those near stream banks, on steep slopes, or areas covered with trees were cultivated. The intense level of agriculture included several very poor farming techniques, such as no rotation of crops, plowing entire fields after each harvest, and allowing animals to graze on stubble and other crop residues. As a result, cattle grazing was reduced, causing more acres to be plowed and planted.

The 1862 Federal Homestead Act also exacerbated land degradation because the farm allotments were too small for the dryness of the western Great Plains. These allotments were later increased and continued to be increased until 1916, although perhaps too late. Many landowners also began to take on more tenant farmers with the increased demand. The percentage of land worked by tenant farmers increased from less than 16 percent in 1880 to more than 40 percent by 1935. These transient tenant farmers had far less commitment to the land than the landowners. They tilled the land excessively by mixing crop residues into the soil, which degraded the soil structure and cohesiveness. They abandoned the land when commodity prices collapsed during the great depression. These depressed prices led to farmers greatly reducing the irrigation of land as well. For example, the Texas Panhandle had 230–250 operating irrigation wells in 1920 but only about 170 wells in 1930.

By the 1930s, the Great Depression had made farming less profitable with the drop in commodity prices, but families still had plenty of food and a relatively good lifestyle compared with the suffering that was occurring in most areas of the United States. However, all of the booming agriculture and prosperity came to a dead halt during the Dust Bowl. There was a catastrophic drop in the amount of precipitation between 1927 and 1934. Areas that normally received 30–40 inches (76.2–101.6 cm) of rainfall annually started receiving 15–20 inches (38–50.8 cm) or less annually. The reason for this drought was a change in ocean currents and the jet stream. This "La Nina" condition in the Pacific Ocean caused maximum drying in the northern and central Great Plains and warm temperatures across most of the United States. It would not be until the 1940s that the precipitation levels recovered.[46]

The drought across the Great Plains was at peak intensity in 1934 and 1935 with average Palmer Drought Severity Indices of -5, -4, 3.5, and 2.8 and the worst from 1870 to at least 2005. The lack of rain was accompanied by extremely hot summer temperatures. In Kansas, 1934 and 1936 were among the top few hottest summers in 180 years. High temperatures were commonly between 105 and 110 °F (40.6–43.3 °C) during the summers, and temperatures above 120 °F (48.9 °C) occurred on occasion. The most intense heat wave ever in the history of the United States occurred in July 1936.[47]

This oppressively hot and dry weather parched the agricultural fields into unconsolidated dust. The plowing had broken the cohesion of the soil structure, and there was no tall grass to bind the soil and hold it in place. As a result, the strong winds of the Great Plains picked up the fertile topsoil and even subsoil and carried it in huge, blowing clouds of black and brown dust. The blowing topsoil dust drifted like black snow, burying farms in layers of soil as much as 6.5 feet (2 m) thick and turning previously productive agricultural fields into barren wastelands. The drifting black dust raised the temperatures by absorbing the hot solar radiation, thereby contributing to the high temperatures. The temperature differentials it created also increased the velocity of the wind.[48] These interacting factors greatly exacerbated the situation, making conditions even worse.

The drifting dust did not just impact the area around the field where it came from. The strong winds whipped it up into huge clouds that were blown for great distances as black dust storms. In 1932, there were fourteen dust storms in the Central Great Plains, and there were thirty-eight storms by 1933. By 1934, the peak of the drought, more than 100 million acres (40.5 million ha) of prime farmland had lost its topsoil to wind erosion. Among the earliest, major dust storms occurred on November 11, 1933. The storm swept through South Dakota, removing all of the topsoil from numerous farms in a single day. The sky stayed dark until noon the next morning across the entire region. On May 9, 1934, strong winds stripped the topsoil from plowed fields and overgrazed pastures in Montana and Wyoming and continued across the Dakotas. It is estimated that the winds stripped up and suspended 300 million tons (272 million mT) of topsoil. The dust storm continued east choking people, destroying crops, and killing livestock and arriving in Buffalo, New York, the next day where it caused total darkness at noon. By the morning of May 11, layers of dust were settling on New York City, Boston, and Washington, DC. This storm finally brought the severity of the situation and plight of the farmers to the attention of people on the East Coast. The situation continued to deteriorate. The following spring, on April 2, 1935, storms stripped up dried fields in Kansas, Texas, Colorado, Oklahoma, and Nebraska. The resulting dust storm darkened the midday skies on Capitol Hill in Washington, DC, as Senate Public Lands Committee was debating the need for a national soil conservation program.[49]

This was just the beginning of the trouble in 1935. On April 10 and 11, wind from the west to northwest at 20 to 30 miles per hour (32–48 kph) raised a dust storm, reducing the visibility to from 60 to 300 feet (18–91 m) in Kansas and eastern Colorado. The turbulence in the dust storm caused static electricity which produced lightning that caused fires. The residents started cleaning the dust out of the houses and off the streets the next day. Most streets had dust an inch thick or more.[50]

The worst dust storm of the entire Dust Bowl was named "Black Sunday," and it took place on April 14, 1935, driven by a strong cold front. By 2:49 p.m., a huge black dust cloud developed in the north. It was 500–600 feet (152–183 m) high and was traveling at 50–60 miles per hour (80–97 kph). The dust was raised high enough in the sky to block out the midday sun, making it as dark as night for much of the day. All traffic came to a standstill. A layer of dust that was inches thick seeped into every house. The dust was so thick that seventeen died in Kansas from dust-induced pneumonia, and three others died from suffocation in the thick dust. The dust storm and accompanying strong winds struck Waynoka, Oklahoma, and Wichita and Emporia, Kansas around 4:00 p.m. Visibilities were similarly near zero with blackout conditions and 35–50 miles per hour (56–80 km) winds. The cold front and dust storm struck Amarillo, Texas with northeast winds of 50 miles per hour (80 km) at 6:50–7:15 p.m. and caused visibilities to reach zero. It then struck Ardmore, Oklahoma, at 10:00 p.m. and Roswell, New Mexico, after midnight. This huge dust storm covered an area more than 800 miles (1,288 km) long and from 300 to 500 miles (483–805 km) wide and extending from central Nebraska to Mexico and Pueblo, Colorado to Eastern Kansas. Up to three million tons of black topsoil blew off the Great Plains and away from the area forever during Black Sunday.[51]

These black dust blizzards ended after 1936 but small dust storms continued until 1940 locally. Exposure to dust was found to carry diseases, as well as damaging the lungs and sinuses. It was determined that the black dust transmitted measles, influenza, and Coccidioides immitis. Mortality rates increased in the areas struck by dust storms. Outbreaks of the diseases occurred with a two- to three-day lag period after the storm.

The human costs as a result of the Dust Bowl were enormous. Once the drought destroyed the crops, most farmers could not afford the mortgage and tax payments on their farms. A huge number of poverty-stricken Midwesterners were forced to abandon their farms and even their homes and leave the area to seek a new life. This resulted in the largest mass migration in American history with more than 2.5 million people, leaving the Dust Bowl states. From 1935 to 1940, at least 440,000 people left Oklahoma, and at least 250,000 of them migrated westward and settled in California, at least one-third in the San Joachim Valley. The novel *The Grapes of Wrath* by John Steinbeck chronicles the hardships of the times and Woody Guthrie sang about it. Many people also left the other Dust Bowl states of Texas, New Mexico, Colorado, Nebraska, and Kansas and settled in other areas. Many parts of the United States became overwhelmed with Dust Bowl refugees. Even though refugees were from all states, they were all referred to as "Okies." The Great Depression had already stressed relief services in these areas, and the refugees overwhelmed them. These newcomers were willing to work for greatly reduced wages, taking work away from local residents. This made them unwelcome in many areas, and there are reports of violence against the refugees.[52]

The response of the local and federal government to the Dust Bowl disaster was very slow. Political leaders slowly came to the realization that much of the productive farmland in the western Great Plains was becoming permanently unproductive. In response, in 1935, the Congress enacted the Soil Conservation Service, which implemented cultivation practices that reduced erosion of topsoil by wind and water. This agency was later renamed as the Natural Resources Conservation Service. New Deal legislation enacted the Works Progress Administration (WPA) and other agencies that provided emergency supplies, funds, feed, and transportation to the farmers and ranchers still remaining on their land. Healthcare facilities were established in these impacted areas, and medical supplies were provided to meet medical needs. The government developed markets for the farm goods, imposed tariffs on imported food products to reduce competition, and offered loans for farm maintenance and rehabilitation. In 1934, the Congressional appropriations for Dust Bowl relief was more than $500 million.[53]

Additional soil conservation policies were implemented later. Water supply and irrigation systems were improved, and a federal crop insurance program was implemented. The size of farms was increased and ecologically sensitive, and buffer lands were no longer farmed. Soil conservation programs were designed to help reduce the impacts of future droughts and to develop national soil conservation practices. Even with all of this legislation and aid, the topsoil has not been replenished in the vast majority of the Dust Bowl area. Most of the soil is red, acidic, and poorly productive at best. It is estimated that the productive topsoil has been removed from about 35 percent of the Corn Belt in the Midwestern states by the Dust Bowl and other poor practices. To replace the topsoil that was lost in just a few years will take thousands of years of natural processes.

Even more concerning is the impact that climate change is having on the Great Plains and especially the Dust Bowl area. In 1936, the summer heatwave was calculated as a 1-in-100-year event. However, with climate change, an equivalent summer heatwave is now considered to be a 1-in-40-year event. In the 2018 US National Climate Assessment, it was predicted that under current warming trends and compared to the 1976–2005 average, temperatures in the southern Great Plains will increase by 3.6–5.1 °F (2–2.8 °C) by 2050 and by 4.4–8.4 °F (2.4–4.7 °C) by 2100. One reason that the Dust Bowl conditions have not returned is that the irrigation of crops was greatly increased. The primary source of the irrigation water is the Ogallala Aquifer, the main component of the High Plains Aquifer System. This aquifer provides water for 30 to 46 percent of the irrigated land in some of the Great Plains states. This aquifer is in danger of collapse as a result of the massive withdrawals and far lower recharge in the past several decades. It has been estimated that the Ogallala Aquifer could be up to 70 percent depleted by 2070. Even with all of the soil conservation and irrigation, if Dust Bowl conditions return, corn and soy production would decline by 40 percent.

9

Floods

River flooding can produce devastating natural disasters that occasionally change history, but it requires particular circumstances. There are two types of river flooding: upstream flooding and downstream flooding. Small streams occur in the upper reaches of rivers. They join together at multiple points to form increasingly larger streams and rivers. The small joining streams are tributaries, and they combine at confluences to get larger, eventually forming a major river. In these streams and rivers, the channel carries the flowing water, and the flat land on either side of the channel is the floodplain. Floodplains are dry most of the time, but they are underwater by up to many feet during floods. Sometimes, rivers shift their channels during floods in flat areas, which can divert the water into unprepared areas, causing disastrous damage. Most of the disastrous floods occur in the major rivers and far from the storm or thaw that supply the water for the flood. Downstream floods can come without warning and cause excessive damage because they can strike when the weather is clear. The reason for the lack of warning is that the storm or thaw occurs in the headwaters of the river, which may be quite far away. The flood in headwaters from the storm or thaw is an upstream flood, and it typically occurs in highland areas in the tributaries. Although these upstream floods can be devastating, many of the upstream floods tend to occur in sparsely populated areas like the mountainous areas. For this reason, there are generally fewer casualties in these events.[1]

Development and urbanization impact the discharge in streams. Vegetation slows the runoff from a rainstorm. Removing it speeds up the runoff reaching the stream. Runoff is quickest if it is flowing over pavement and buildings. They also reduce the amount of infiltration into the soil during precipitation. The combined effects of speed and volume of runoff caused by development make the size and timing of floods much more devastating than they would be naturally. For this reason, flash flooding is greatly increased in developed areas.

Coastal flooding typically results from major storms with tsunamis a distant second. However, their time of arrival in the tidal cycle can greatly exacerbate the impact of the event. Tides are the regular rise and fall of the water surface of the ocean caused by the response to the gravity primarily of the moon but also from the sun. Oceans typically have two high tides and two low tides per day separated by twelve hours. The high tide occurs when the moon is either directly overhead or directly over the opposite side of the earth. The low tide occurs when the moon is on the horizon. If a high tide is at midnight, there is a low tide at 6:00 a.m., a second high tide at noon, and a second low tide at 6:00 p.m. If storm surges or tsunamis strike the coast during high tide, their heights are additive, and they are much more damaging than if they strike at any other time but especially at low tide.[2]

The height of the ocean surface between high tides and low tides is the tidal range. The distance that the tides penetrate inland during high tide is the tidal reach. The ocean height of high tides and low tides, and therefore the tidal reach, varies throughout the month because of the varying effects of the gravity of the sun. When the sun, moon, and earth align in any configuration, the high tides are at their highest, and low tides are at their lowest. The reason for these height extremes is that the gravity of the sun and the moon is additive in this configuration. This is called a spring tide, and it corresponds to the full moon and the new moon. If the sun and the earth make a right angle with the moon and the earth, there will be a neap tide. Neap tides are characterized by lower high tides and higher low tides than normal. The distance between the earth and the moon and the earth and the sun also varies. This has a lower impact than the alignment, but it can have a small effect. If a storm or tsunami strikes at high tide during spring tide, it will cause the most damage.

However, the most controlling factor in the tidal height, range, and possibly reach is the shape of the ocean basin that it impacts. Small, shallow ocean basins produce a smaller tidal range and reach, and only have one tidal cycle per day in some cases. Large, deep basins produce bigger tides. The bathymetry of the seafloor and the shape of the shoreline also play a big role in tidal range and reach. The tidal range in New Jersey averages about 6 feet (1.85 m), but in the Bay of Fundy, Nova Scotia, Canada, the tidal range is 32 feet (9.85 m), although both face onto the same coast of the Atlantic Ocean. The difference between them is the shape of the coastline and that the ocean floor is shallow and flat off the New Jersey coast but steep off of Nova Scotia. All of these factors increase the impact of coastal flooding.

1287 St. Lucias Flood

During the last ice age, the sea level dropped by about 200 feet (67 m), and many areas of seafloor became lowlands. One of these areas was the span between Europe and Great Britain. Great Britain was part of the continent of Europe rather than an island like it is today. This area is called Doggerland, and it was nearly 18,000 square miles (46,620 km^2), and it was inhabited by humans in addition to allowing people and animals to cross between the two regions.[3] However, about 12,000 years ago, the ice age ended and the continental ice sheets began to melt. This gradually raised the sea level to current levels. However, the weight of the ice had depressed the land surface, and it slowly arose to the levels of today. Therefore, there was a battle between the rising sea level and the land surface that complicated the flooding of Doggerland.

Finally, about 6,000 years ago, the flooding prevailed, and the Mesolithic inhabitants were progressively forced onto higher ground in Great Britain and the Netherlands. This flooding slowly formed the English Channel and part of the North Sea.

Whereas Great Britain has relatively small areas of lowlands and ample highlands, the Netherlands are low-lying compared to the neighboring North Sea.[4] In fact, "Netherland" means lowlands. This area was largely a tidal wetland composed of mudflats and peatbogs with tidal waters washing in and out around them.[5] The sea level did not rise gradually but instead in a series of several uneven bursts. These events involved erosion of the south side of the coast, withdrawal to the north, and then re-advancement to the south. These cycles eroded the land near the sea, and the bays were washed out. These cycles left inhabitants of the coast rushing from one area to another in an attempt to cope with the shifting of inhabitable land.[6] Residents attempted to build mounds around peat bogs for settlements, but these were marginally successful. By about 800 CE, residents changed their approach to building dikes or walls to prevent flooding and sluices and canals to drain and remove unwanted water. As a result, they were able to farm the land and build more permanent towns. Drained and reclaimed areas were called "polders." However, during the ice age, much of the area had been filled with thick layers of peat, which were 11.5–26 feet (3.5–8 m) thick below the ground surface. Between water removal from wetland for farming and the widespread peat-bog cultivation and harvesting, the area began to subside significantly. It is estimated that, on average, the land surface in the Netherlands subsided about 3 feet (1 m) per century. This left a substantial part of the Netherlands below sea level and therefore vulnerable to storm surges from the North Sea.[7]

This land reclamation spread across the south shore of the North Sea, across Frisia, and into Germany. Much of the building of dikes and drainage of land in the area occurred in the eleventh and twelveth centuries CE by local residents or associations/cooperatives of residents. At first, the dikes were not built high enough to protect against storm surges, but soon the residents felt well protected. As a result, more salt marshes were inhabited, and the landward lowlands were drained, especially across the Netherlands and in the northern part of Eiderstedt in Schleswig-Holstein in Frisia.[8] This left small-to-moderate settlements built on slightly elevated areas of sand between vast agricultural lands that were below sea level all along the North Sea coast.

While all of this draining, reclaiming, building of dikes, and establishment of towns and villages was occurring, the climate was in the Medieval Warm Period. This climate produced calmer weather and fewer, less powerful storms. The inhabitants of the North Sea coast flourished during this time and built a society in this marginal area based on the climate they had become accustomed to. This Medieval Warm Period lasted through the twelfth century but began to become unstable after that. The climate then transitioned into the Little Ice Age, which involved significant cooling and an increase in very powerful storms.[9] In contrast to the earlier period, the thirteenth through fifteenth centuries were marked by regular devastating and disastrous storms in the North Sea region, but none as history changing as the 1287 St. Lucia's storm and flood.

Virtually all of the catastrophic storm floods in the North Sea region were generated by large storms with extreme north-westerly gales that occurred during spring tides. Spring tides occur when the sun, earth, and moon line up, thereby amplifying the gravitational force on the surface water. This generates the highest high tides. Combining these, elevated sea levels with the surge from a major storm can prove disastrous. This kind of storm struck the North Sea area on December 13–14, 1287 CE as the residents of the area celebrated the Feast of St. Lucia.[10] The elevated storm surge struck the entire North Sea coast, breaching the dikes and seawalls along the north coast of the Netherlands. The coastal dunes and sandbanks did not withstand the surge and were completely washed away. The low-lying agricultural plains were quickly submerged, and many towns and villages were swept away. This included the major commerce ports along the North Sea like powerful trading port of Stavoren; they no longer existed. Immense areas of the coast including villages and cities were completely submerged, some permanently. All of this destruction is also estimated to have caused the deaths of between 50,000 and 80,000 people, but considering that whole towns were removed, much less individuals, accurate numbers are very unlikely. In any event, a large percentage of the residents of northern Netherlands, Frisia, and part of Germany perished. More than 500 people in England also perished.[11]

The geography of the North Sea coast was completely changed by the flood, and in many cases, it was permanent. The water that did not recede created several bays, lakes, and islands that still remain. The few remaining settlements that were previously inland wound up as coastal towns and cities or were on islands surrounded by water with no means to reach land. The biggest change, however, was in a small lake that was drained by the River Vlie to the North Sea. The river was destroyed and the lake scooped out, forming a shallow bay that that still extends 60 miles (100 km) inland as the Zuiderzee or Southern Sea. A small previously inland settlement now sat in the ideal place on the Zuiderzee to serve as the major port for the region. This settlement called Amsterdam quickly grew into a major city and the trade hub for the region. It would eventually serve as the capital of the Netherlands.[12] This flood certainly changed the history of the Netherlands, but Amsterdam played a major role in European trade for many centuries, and it continues to be important today. Without the flood, the history of Europe would not have been the same.

The Netherlands was not the only country that suffered from the flooding. Coastal Frisia was at least as badly damaged. The coastline shifted many miles inland, making previously landlocked cities, like Harlingen, into prosperous coastal ports. Previously inhabited islands in the North Sea, like Griend, were wiped off the map or nearly so. Part of coastal Germany similarly had the coastline completely changed, and entire towns and cities were lost there as well. The death toll from these areas is estimated to be included in the 50,000–80,000 but it could be much greater.

The Brittany coast was also impacted by the St. Lucia Flood. It heavily eroded marshland in northern France and left deposits of gravel and coarse sand,[13] although there are few reports of human impact. The storm and flood also struck England's coast, greatly impacting its geography as well. The prosperous port city of Winchelsea was completely destroyed and abandoned, as was Broomhill.[14] A cliff collapsed at Hastings, blocking the harbor and destroying much of a Norman Castle. The storm surge silted up the harbor by New Romney so heavily that it diverted the city's river at least 15 miles (24 km) to the west, turning the former port city into a landlocked town. The heavy silt elevated the surface of the town by 5 inches (13 cm). At least 180 residents were killed by flooding in the village of Hickling. In all, at least 500 people died in Great Britain including Hastings, Romney, and Winchelsea. The major port and once Anglo-Saxon capital, Dunwich, had a population in excess of 3,000 people but was so damaged by the storm and flood that it never recovered. As a result, it was gradually abandoned over time, leaving a small village of 180 people.[15]

1927 Great Mississippi River Flood

The Emancipation Proclamation was the most important and powerful legislation in the young American nation. At the time, about four million slaves were freed from their owners as a result. Because it was wartime, the implications did not become apparent for a few years. The southern economy was almost exclusively agrarian at the time, and the massive destruction from the war prevented any transition to a more advanced mode. Prior to the war, landowners could use their slaves as collateral to secure credit from merchants to purchase seeds, supplies, and farming equipment. The emancipation of the slaves without compensation devastated the net worth of the dependent landowners, meaning that they had much less collateral. Some freed slaves were given plots of land after the war ended, but they were not given instruction on farming techniques or the economics of managing a farm. Many lost their land. The result was large numbers of recently freed slaves in need of work and money on one hand and plantation owners who were unable to farm their land and without enough money to purchase supplies and hire workers to farm them on the other hand. Borne out of these contrasting needs was the new solution of sharecropping.[16]

In this arrangement, landowners carved up their property into plots and leased them to families or groups to farm them. The family took on the burden of cost, farmed the land with guidance from the landowner, and then shared in the profits of the harvest. Depending upon the arrangement, the sharecropper took on the debt and paid it off. Southern landowners began to depend on sharecroppers and cheap Black labor to run their own operations in the post-war period.

This was an acceptable situation for the southern landowners, but it was so unreliable that they essentially reinstalled slavery under another name. Because they owned no farming tools, sharecroppers were forced to obtain credit from plantation stores to purchase needed food, clothing, and tools at highly inflated prices. At harvest, the credit accounts were paid off, leaving the sharecropper with little to no money. As a result, they were commonly forced to work an additional year under the same arrangement because the landowner kept the indebtedness so high. In the Mississippi lowlands, Black people were 75 percent of the population yet furnished 95 percent of the plantation labor. By requiring poll taxes and literacy tests, most Southern states essentially disenfranchised the entire Black population. They had no political power to change the situation at all. To keep this new arrangement, Southern white landowners used their political power to restrict the mobility of Black laborers in general but especially Black farm workers. These landowners could not survive without cheap Black labor. As a result, new laws were enacted, making it illegal for one landowner to hire another landowner's workers. At the same time, new anti-vagrancy laws made it illegal for people to be unemployed and without housing, thereby trapping the Black laborers in an inescapable net. As a result, the Southern economy remained agrarian and underdeveloped between the Civil War and the Second World War.[17]

This situation persisted from the Civil War until after the First World War when the Southern Black farm laborers began migrating in massive numbers from South to North, especially to the northern cities like Chicago. This put pressure on the southern landowners to compete for cheap labor. Part of the pressure to force the Black laborers to remain in the South was an explosion of Ku Klux Klan activity in the 1920s, brutalizing and terrorizing many thousands of African Americans to scare them into remaining.[18] Then, in the mid-1920s, cotton prices fell sharply, further compounding the problem. It almost bankrupted plantation owners who were forced to mortgage their properties and take bank loans on an annual basis. But in order to secure the loans, they were required to prove that they had one tenant family for every 20 acres. With the dwindling labor supply resulting from fleeing Black workers, this became more difficult and escalated the tensions.[19]

The stage was set for the perfect storm, which only needed a small amount of encouragement. However, it was the natural disaster of the 1927 Great Mississippi River flood that caused the true change in history. The disaster began innocently with heavy rain over the upper Midwest around the middle of August in 1926. This first storm lasted many days, first in South Dakota, Nebraska, Kansas, and Oklahoma and later in Iowa, Missouri, Illinois, Indiana, Kentucky, and Ohio. Just two days later, a second storm and then a third storm struck the Mississippi River Valley with rain lasting two weeks. By September 1, many streams and rivers were overflowing and flooding towns in the flood plains. The rain continued throughout September and into October, causing the Mississippi River to rise rapidly and wash out several bridges. By the end of

October, the continuous rain finally stopped, but flooding had already crossed the entire upper Midwest. The pause in the rains, however, was short-lived as it started raining again by mid-December across the Mississippi Valley, and snow fell across the northern states. By Christmas, flooding had already made thousands of residents homeless, and railroad traffic across the Mississippi River was suspended. The flooding caused the Ohio, Missouri, and Mississippi rivers to be at the highest levels ever recorded from October to December of 1926.[20]

The solution to controlling the frequent floods in the Mississippi River was to build an extensive system of levees. The Mississippi River Commission was established in 1879 to oversee this project. Levees were built along 1,100 miles (1,770 km) of the lower Mississippi River from Cairo, Illinois, to the Gulf of Mexico. It was soon apparent that the system was not built for the flood of 1927 and that action would need to be taken as the rains were falling again by mid-January 1927. On January 28, 1927, the Ohio River flooded the city of Cincinnati, and in early February, the Little Red and White rivers flooded more than 100,000 acres in Arkansas. In early March, heavy rains continued to drench the lower Mississippi Valley, while to the North, snow fell from the Rockies to the Ozarks. By the end of March, every levee commission on the Mississippi River south of Cairo, Illinois, was patrolling the levees for breaches twenty-four hours a day, seven days per week, and sandbagging the levee tops to prevent the floodwaters from overtopping them. This was an effort in futility because by the mid-April, more than 1 million acres of land were underwater, and at least 50,000 people were homeless and living in temporary shelters in Arkansas, Illinois, Kentucky, Mississippi, Missouri, Oklahoma, and Tennessee. By this point, the entire Mississippi River from Iowa to Louisiana was in full flood stage.

On Good Friday on April 15, 1927, some 15 inches (38 cm) of rain fell on New Orleans, Louisiana, over just eighteen hours from a simple rainstorm. This was one-quarter of the annual average rainfall for the city, and the record exceeded even rainfall from hurricanes. The same storm dumped more than 8 inches (20 cm) of rain in Greenville, Mississippi, and 6–15 inches (15–38 cm) fell across the lower Mississippi River valley. This was an immense amount of precipitation to add to the saturated ground and rivers already at the top of the bank if not actively flooding. The public began to panic, and although the US Army Corps of Engineers assured them that the levees would withstand this new onslaught, they did not.[21]

On April 16, a 1,200-foot (366 m) section of Mississippi River levee collapsed at Dorena, Missouri, just 30 miles (48 km) south of Cairo, Illinois. The breach released enough river water to deeply flood 175,000 acres (70,820 ha) of land. The breach was termed a crevasse, which is basically a small delta that forms in the floodplain from a breach in an overflowing river. In response, thousands of men were immediately conscripted to reinforce the levee with sandbags along the lower Mississippi River twenty-four hours per day under grueling conditions. Despite these efforts, the next section of levee to break was at Mounds Landing, just below the confluence of the Arkansas River about 12 miles (19 km) north of Greenville, Mississippi. This breach occurred on Thursday, April 21, at 8:00 a.m. The breach and resulting crevasse were more than 3,960 feet (1,207 m) long and 100 feet (30 m) deep, allowing phenomenal amounts of river water to rush across the land. The city of Greenville was completely flooded within one day. In just ten days, an area of more than 1 million acres (405,000 ha) of land across the Mississippi Delta region was under at least 10 feet (3 m) of Mississippi River water. During the remainder of April and the whole month of May 1927, multiple levee breaches occurred all along the river to the South. The final natural breach of the levee system occurred on May 24 at McCrea, Louisiana, on the east bank of the Atchafalaya levee.[22] In all, thirteen major crevasses developed in addition to many smaller ones. Reports claimed that it appeared as if the levees were crumbling all along their length.

Much of the city of New Orleans sits below sea level and far below the level of the Mississippi River that flows through it. With the regular news of levee breaks and massive flooding progressing downstream and ever closer dominating the radio broadcasts, the residents of New Orleans began to panic that their city would soon be profoundly flooded with many casualties as a result. Under these circumstances, it was not difficult for the city fathers to convince the residents that drastic action needed to be taken. This drastic action was to dynamite the levee at Caernarvon, Louisiana, downriver of the city on Friday, April 29. The goal of the blasting was to release the swollen and backed-up water in the Mississippi River as it passed New Orleans, thereby reducing the height. However, the levee breach at Caernarvon flooded much of Plaquemines and St. Bernard parishes including the houses and fields of the poor people living there. The general opinion of the residents of New Orleans was that the area was sacrificed for the greater good of the region. In fact, the severe Mississippi River flooding and levee breaks never reached as far South as New Orleans, so the blasting and sacrifice was unnecessary.[23]

By the peak of the 1927 Great Mississippi Flood, at least 27,000 square miles (70,000 km^2) of the lower Mississippi River valley inundated with flood waters up to 30 feet (9 m) deep. This area of flooding is equal to the area of Massachusetts, Connecticut, New Hampshire, and Vermont combined. The river was 70 miles (113 km) wide at that time. It would take until mid-August of 1927 for the waters to finally recede to normal conditions. However, it left a barren, mud-covered landscape as far as the eye could see. It was estimated that, in all, more than 700,000 people were left homeless, and of these, approximately 555,000 were racial or ethnic minorities. In addition, at least 330,000 people had to be rescued from rooftops and placed in refugee camps on levee tops and other minimally high grounds. The floods damaged or destroyed at least 137,000 buildings and caused the deaths of at least 250 people and probably closer to 1,000, as African Americans were not included in the lower number. Direct property losses had an estimated value of \$250–500 million, but with the indirect losses, the actual costs were closer to \$1 billion.

This flood certainly was a severe natural disaster on its own. However, the reason that it changed history was the response and actions of the White residents and government officials with regard to the Black residents. Not that the Black residents were treated much better than they had been when they were slaves prior to the flood, but during and after the flood, their treatment was appalling and that changed the history of the United States. The problems began before the flood breached the levees. When the levee commissions attempted to prevent the overtopping of the levees, they set up what amounted to concentration camps on the levee, especially to protect Greenville, Mississippi. Thousands of African American plantation workers were drafted to raise the height of the levee by filling and stacking sandbags on the top. These camps included field kitchens and tents for the men to live. However, the work effort was around the clock with White supervisors using brutal tactics to get as much work out of the Black workers as they could, and if they attempted to escape, they were beaten and returned to the camp.[24]

Despite the efforts, the levee was breached on April 21, 1927, and a huge 100-foot (30 m) deep, one-half mile (0.8 km) wide channel or crevasse opened just north of Greenville. Water poured through this gap at twice the flow of Niagara Falls and covered one million acres at 10 feet (3 m) depth within ten days. The White plantation owners panicked that if the Black farm laborers fled the area, they would never return, and there would be no one to tend the farms. To prevent this catastrophe, the White landowners coerced the government officials to order all Black people from the Greenville area to proceed to the levee. There, they set up a camp that stretched 7 miles (11.3 km). The Red Cross was enlisted to run the this, and the other similar camps on the remaining levees. However, there were not enough tents, insufficient food supplies, no eating utensils, and no mess hall for those in the Greenville camp, and many other camps for Black residents had similarly reprehensible conditions.

When President Coolidge was confronted with news of the flooding and plight of the refugees, he showed concern but did nothing. Instead, he assigned the situation to the then Secretary of Commerce, Herbert Hoover, who chaired a special emergency committee that handled operations dealing with the flood. He effectively used this high-level position not to alleviate the suffering of the refugees but instead to gain extensive publicity for himself. Through regular media throughout the disaster, he claimed great success in dealing with the situation and saving the lives of many Americans, although his actions were minimal. As a result, he became a hero and through it, he won the 1928 Republican nomination for president and subsequently was elected the 31st president of the United States.

Instead of the rosy account painted by the future president, the situation in the relief camps was grim for all and terrible for most. In all, 154 relief camps were established by the National Guard, the Department of Commerce, and the Red Cross. Of the 608,000 homeless people in the relief camps, 555,000 were Black agricultural workers, and the camps were segregated. However, all camps were almost exclusively administered by Southern white people. In White relief camps, the National Guard aided in camp administration but were mostly advisors to the refugees. In contrast, relief camps for Black refugees served as holding pens to ensure they would not flee the area and deplete the local agricultural labor force. The camps were guarded by armed National Guard soldiers who prevented refugees from leaving and any outside people from entering, all at gunpoint. The Black relief camps further lacked in many supplies from tents to eating utensils, and the people were purposefully supplied with inferior food in comparison with the White camps.[25]

As a further insult, as part of the camp registration process, the Red Cross administrators tracked who each of the Black farm laborers worked for before the flood, but no such tracking was kept for White refugees. When the flood waters receded, the White plantation owners were allowed into the Black refugee camps where they legally reclaimed their workers. If the Black workers refused to go with the plantation owners, they were required to provide concrete proof that they had an alternate form of support, or they were forced to return to the plantation at gunpoint. This proof was required to be submitted in the form of a written letter. However, most of the Black farm laborers were uneducated, and most were unable to read or write. Further, their Red Cross support was cut off once the plantation owner came to retrieve them. In addition, there were labor agents who attempted to enter the refugee camps to offer Black refugees alternative employment opportunities on other plantations, in other states, or even in the North at other, more lucrative jobs. However, these agents were barred from entering relief camps by the National Guard at gunpoint. To make sure they would be unsuccessful, White landowners organized themselves into groups to prevent the labor agents from attempting to entice Black refugees from leaving the camps, commonly through apprehension, threats, and violence. There was a more formal name to this practice called the "Yazoo Plan," which supposedly ensured that labor agents would not be harmed upon being apprehended, but this was rarely the case.

The flood permanently further strained relations between Black and White residents of the southern Mississippi River region. Several incidents further inflamed the situation. A White police officer shot and killed a Black man because he refused to unload a relief boat. Another Black man was lynched after killing a plantation foreman. He had been trying to force the Black man to rescue mules that belonged to the plantation owner. As a result of the poor treatment and despite the best efforts of White plantation owners to bully the Black laborers, the counties that had been flooded experienced an immediate but continuing migration of the Black residents. This depleted the manual labor force of the area and ultimately forced the landowners to modernize their agricultural production.[26]

The impact of the 1927 Mississippi River flood on population demographics in the United States was that it greatly accelerated the Great Migration of African Americans from the South to the North. After the

Emancipation Proclamation and the freeing of the slaves, most African Americans remained in the South. As of 1910, about 90 percent of African Americans lived in the South. The migration out of the South began soon after but increased after the First World War when higher-paying jobs and a shortage of labor began in northern states.[27] As a result, tens of thousands of African Americans relocated to the upper Midwest's big cities like Chicago. The 1927 flood greatly accelerated the migration. In fact, the 1920s had a much higher rate of migration than the 1910s and 1930s as a direct result of the flood. It would not be until after the Second World War that this rate would be exceeded.

The flood became a part of popular culture both while it was occurring and ever since. Journalist Harris Dickson wrote a twelve-part series on the flood that was nationally syndicated. Many prominent people donated money to the refugees, and several foundations raised significant funds. The popular film star Will Rogers raised more money than anyone to help flood victims. The flood found its way into music in the Mississippi Blues including "Mississippi Heavy Water Blues" by Barbecue Bob in 1927, "Backwater Blues" by Bessie Smith in 1927, and "When the Levee Breaks" by Kansas Joe McCoy and Memphis Minnie in 1929. A cover of this last song was reworked by the rock group Led Zeppelin and released in 1971 for their fourth album. The great bluesman Muddy Waters also referred to the flood in several of his songs but well after it. The book "Lanterns on the Levee: Recollections of a Planter's Son" (1941) by William Alexander Percy described the flood in great detail. William Faulkner described the flood in his novel *The Wild Palms*, which was later retitled *If I Forget Thee, Jerusalem*. The description was originally a novella entitled *Old Man*. William Faulkner further described the flood in his novel *As I Lay Dying*, released in 1930. In it, a character describes the flood as "[you] couldn't tell where was the river and where the land. It was just a tangle of yellow and the levee not less wider than a knife-back." Richard Wright also included the flood and its aftermath in his writings.[28]

These portrayals were not just descriptions of a tragedy. They were largely direct criticisms of the society that would allow such human inequities to occur. They showed that the flood made it very clear that human progress since the Civil War had not been equitable across races or even regions. Further, what had developed was clearly not sustainable. These writers expressed a general sentiment of much of the country that the residents of the lower Mississippi River in 1927, including the White plantation owners, but especially the Black and poor white laborers, were historically retrograde. They were trapped in feudal social and labor relations and outmoded Christian fundamentalism. In 1925, the northward migration of southern Blacks was described as a temporal leap forward "from Medieval America to modern."

Perhaps the only positive outcome of the Great Flood was the quick proposal and passage of the Flood Control Act of 1928, which was signed into law by President Coolidge on May 15, 1928. This legislation gave the federal government authority to contain and regulate the lower Mississippi River. As a result, significant federal funding has been dedicated to ensuring that the flood of 1927 would not be repeated. The act and its enforcement gave the Army Corps of Engineers the authority to build an impenetrable levee system. There have been several instances since 1928 when the system was severely tested, but it has certainly proven to be much more formidable than its predecessor.[29]

1938 Yellow River Flood, China

It takes the involvement of humans to turn a natural geological event into a natural disaster. In numerous cases, human actions and reactions can exacerbate the disaster, making it much worse than it would have been. But there are some natural disasters that had so much human involvement that they might be better classified as unnatural disasters. This was the case with the 1938 Yellow River flood. This flood was directly caused by the removal of structures to prevent flooding, namely man-made levees. The result was a terrible flood that changed both Chinese and world history.

There have been many devastating floods in China. The 1931 Yangtze–Huai River flood was apocalyptic, completely submerging major cities, leading to epidemics from waterborne diseases and resulting in the deaths of up to 4 million people. But the government of China stayed the same, and some countries and individuals aided the victims with no other outcomes. The 1975 Banqiao Dam collapse and flood resulted from once-in-2,000 year flood is considered one of the top 10 technological disasters in history, and it killed about 171,000 people including victims of diseases, but there were no changes in China, much less the world.[30] In contrast, the 1938 Yellow River flood not only killed and displaced huge numbers of people but it also impacted the outcome of the Second World War and helped spark the People's Revolution and the conversion of China into a communist country.

The Yellow River flows for 3,395 miles (5,464 km) across the mainland of China, making it the second longest river in China and one of the most important in eastern Asia. Blowing loess from the arid areas is deposited in the water where it is carried as silt in extreme amounts. The silt gives the water a yellowish, turbid hue for which the river is named. The headwaters of the river are in the southern Qinghai province on the Tibetan Plateau. The steep topography there and throughout the middle and lower courses holds the river tightly in its channel. However, in the lower course beginning just downstream from Zhengzhou, the Yellow River widens as it flows across the featureless North China Plain. The topography is so flat through Henan and Shandong provinces that the river regularly shifted its course, causing devastating floods. In an attempt to keep the river in a single channel, the Chinese have been building levees, digging channels, and building dams since the third century BCE. The problem is that deposition in the channel makes the river higher than the surrounding countryside. Breaking the man-made flood controls happened regularly, and any time the river was exceptionally high. These devastating floods continued until modern times with major floods in 1887, 1889, 1921, and 1933 before the flood in 1938. The Yellow River empties into the Yellow Sea and forms a large delta. Shifting channels cause devastating floods there as well and has shifted settlements throughout Chinese history.[31]

During the middle to late 1920s, imperial Japan became increasingly militaristic. By 1930, they decided that they required a foothold on Mainland Asia for economic purposes and began to flex their military might. For this reason, they invaded and captured part of Manchuria in 1931. The Chinese leader, Chiang Kai-shek, dealt with these invaders through appeasement and compromise and instead devoted his efforts to destroying the fledgling communist party. Finally, his own troops kidnapped him and forced him to drop those efforts and confront the ever-aggressive Japanese. Reluctantly, he joined with the communists to militarily confront the Japanese who were threatening their northeast border. As a result of this new stance, in July 1937, a minor skirmish at Marco Polo Bridge near Beijing escalated into the Second Sino-Japanese War, which lasted from 1937 to 1945.[32]

This minor pushback was all that Japan was waiting for. They launched a full-scale military offensive southward into the heart of the Chinese empire. The Chinese fought valiantly against the militarily superior Japanese army for about five months in a series of horrific battles, but Japan clearly dominated and pushed ever forward. By December 1937, the Japanese seized the city of Nanjing, the capital of Nationalist China. The invading soldiers committed brutal atrocities to the civilian residents and the event became known as the Nanjing massacre. The government scrambled to relocate the capital to the city of Wuhan with the Japanese right on their tail. In early 1938, the Japanese launched twin assaults along the Jin-Pu railway. The two joined in Xuzhou and headed west toward Zhengzhou in Henan, a major railway junction that controlled supply transport for the region. If the Japanese took the railway, they would be able to bring in troops by train. After that, they would take the new capital of Wuhan.

By this time, the Chinese army was in disarray. The Japanese army advanced at will with little to no resistance. By May 1938, only two of China's major cities were not controlled by the Japanese. The imminent surrender at Zhengzhou and Wuhan was only days away when, in utter desperation, Chiang Kai-shek made a fateful decision. On June 4, 1938, he ordered Chinese troops to breach the levee along the Yellow River and flood the countryside. His plan was to use flood water against the Japanese troops to stop their advance and inflict as many casualties as possible. However, destroying the levee was not as easy as he thought. Large numbers of Chinese soldiers attempted to breach the levee in three places over the next six days. They used explosives, shovels, chisels, and hammers, and on June 9, after two failed attempts, the river flowed through the hole at Huayuankou, a small village just north of Zhengzhou and just 30 miles (50 km) west of the Japanese army.[33]

Although the flood started out slowly at first, it had the desired effect. The Japanese advance came to an immediate halt with leaders not knowing what to do. As the river flooding increased, so did the flood across the region. It killed numerous Japanese soldiers, prevented large-scale troop movement, and mired down tanks and mobile artillery so that they could not be used. Erosion of the breach in the levee widened into a 5,000-foot (1,525 m)-wide channel that captured the entire Yellow River and causing it to abandon the course it had been following since 1855. The river now flowed southeast rather than northeast, and it flooded much of the Henan Province. It joined with the Huai River in the Anhui Province and flooded most of north Anhui before flooding northern Jiangsu Province. This flood killed and displaced huge numbers of Chinese residents of these areas.[34]

Although effective at first, unfortunately, the flood only postponed the Japanese attack on Wuhan by five months instead of stopping it. The city fell in October 1938 but not before the Chinese government retreated to the west of the flooded area and outside of Japan's reach. Without the flood, the government would most certainly have fallen within days. It also halted the

Japanese from attacking Zhengzhou and the areas west of the flood in Henan and Anhui until 1944. It allowed the Chinese to maintain control of a 150-mile (241 km) section of the main north-south railroad in the area, which prevented the Japanese from transporting supplies to their troops by train. It also allowed the Chinese troops to maintain control of all areas west of the river for many years with minimal engagement by the Japanese. It would take until April 1944 for Japan to launch their Ichigo Offensive and finally take full control of China, however temporarily.

The 1938 Yellow River flood changed the course of the Sino-Japanese War and the Second World War. Without it, Japan would have taken China in 1937, removed all weapons from the Chinese, and installed a wartime government under martial law. They would have been able to put minimal resources into China and instead focused on their other conquests around the Pacific and in Asia. But because they were still at war in China, they had to maintain combat-ready troops, munitions, and a guarded infrastructure all in the middle of a major natural disaster. This stalemate lasted seven years while the Japanese military was stretched along several fronts fighting formidable opponents. By the time they finally managed to defeat the Chinese, the other fronts were all closing in on them, and the Second World War was beginning to conclude. The 1938 Yellow River flood therefore gave China their freedom for seven more years and weakened Japan in the Second World War, limiting their conquests and hastening their demise.

However, these were not the only changes in history caused by the 1938 Yellow River flood. The flooding may have been effective at slowing the Japanese army long enough to keep the struggling Chinese army government from defeat but it also wreaked havoc with the residents of the flooded areas. The extreme human suffering it caused greatly outweighed the benefits to the army, and the government would pay for their decision in the end.

It is estimated that the floodwaters covered an enormous 27,000 square miles (70,000 km^2) of largely farmland southeast of the Yellow River breach and kept it out of full productivity until 1947. The flood was not just in 1938, but every year when the river flow peaked, new floods would engulf the countryside, and with the shifting channels, no predictions could be made about what would be safe and what wouldn't. In the agricultural area of eastern Henan, the floodwaters inundated at least 32 percent of the farmland and buried it in approximately 100,000,000 tons (91 million mT) of silt, which left it an unproductive wasteland for many years. In Henan Province, it is estimated that 325,589 people perished in the flooding and 1,172,687 were displaced, some 900,000 in 1938 alone. In the whole flood-impacted area, the number of refugees topped 4 million.[35]

This instability and lack of farmland badly damaged productivity. Upon the start of the war in 1937, grain production in Henan Province decreased by 30 percent from 1936. The Yellow River flood in 1938 decreased grain production to less than 50 percent from 1936. This resulted in widespread food shortages. If conditions weren't bad enough, a strong El Niño event struck east Asia between 1939 and 1942. As a result, Henan Province and nearby areas of North China experienced a sharp decline in precipitation. The drought was exceptionally bad in Henan in 1941–1942, causing grain production in the area to decrease to just 22 percent of the 1936 output. Adding to the drought and damaged fields, the dried Yellow River flood area encouraged locust reproduction, and the locust swarms further reduced the crop output. The destruction of infrastructure caused by flooding, serious drought, and seizures for the one million soldiers in the area all contributed to the great Henan famine of 1942–3. This famine led to the death of between two and three million people. Included in the horrors faced by the refugees from these events, they also suffered through outbreaks of malaria, cholera, dysentery, smallpox, measles, jaundice, typhoid, scabies, and skin ulcers. This added to the death toll. There are also reports of bands of criminals raiding towns to steal food and resources. There was one case where more than 700 raiders invaded a village and burned more than 160 houses before the authorities could come to their rescue. Compounding all of these deadly factors, as many as 900,000 people perished from the 1938 Yellow River flood.[36]

When the levees were breached, only the people directly around the river were informed. Everyone else was taken by complete surprise so were totally unprepared. The Chinese government announced that the breach was caused by Japanese aerial bombing during the battle rather than the actual military tactic by the Chinese. Even though the Japanese vehemently denied the accusation, the Chinese newspapers and even the Communist party did not contradict the report. As the war continued, the Chinese government declared that the flooding was a national sacrifice to help protect the rest of the country. However, as the impacted area kept expanding and the suffering of the residents of these areas grew worse, the Chinese Communist officials and newspapers split with government propaganda. By 1946, they publicly charged the Chinese Nationalists with having practiced a cruel and outdated form of warfare when they breached the levee with no thought for the hardship they would inflict. This stance gained favor with the impacted residents, and as a result, the Communists gained control over large parts of the flooded areas in the provinces of Henan, Anhui, and Jiangsu. By the end of the war, they were a formidable force in the region.

As early as 1939, the Chinese Nationalists had drafted hundreds of thousands of residents in and around the flooded areas not only to make several feeble attempts to repair the old levees but also to build new levees around the shifted river course. This was very dangerous work because the river was commonly the boundary between the Chinese and Japanese troops. However, none of these efforts worked, and by May 1943, there were seventeen breaks in the levees in three counties. The United Nations Relief and Rehabilitation Administration (UNRRA) was established to address severe and dangerous situations in 1943. Immediately, the Chinese Nationalists requested hundreds of millions of dollars in aid to repair the levees. They convinced the UNRRA officials that diverting the river back to its original course and repairing the levees would reclaim millions of acres of productive farmland and relieve the plight of the residents. This rush to restore the river, however, developed into a major point of contention between the Nationalists and the Communists. The Communists argued that some 500,000 people had moved into the old Yellow Riverbed during and after the war and were farming it. They would need time to be relocated. By January 1947, the Communists and Nationalists were in a stalemate over the issue, and the Communists were blaming the Nationalists for all the damage that the breach had caused. Despite the disagreement, the Nationalists worked out an arrangement with the UNRRA to employ 330,000 Chinese laborers and using UNRRA equipment for six months to plug the breach and repair 400 miles (644 km) of levees along the original river course and return the river to its old channel. On March 15, 1947, they succeeded in closing the breach and held a huge celebration in an attempt to humiliate the Communists.

Two years later, the Communist Party launched the People's Revolution with exceptional support from the residents of the flooded area. They felt abandoned by the Nationalist government and supported by the Communists. With this tremendous support, the Communists swept across the country and banished Chiang Kai-shek to Taiwan. The 1938 Yellow River flood wasn't the only issue in the revolution and radical change in the history of China and the global Cold War at the time, but it played an important role.

10

Can It Happen Again?

If a natural disaster strikes just the right place at just the right time, it can change history. This may be local history, national history, or global history depending on the time, location, and severity of the event. The immediate event certainly causes a disaster and is never positive. However, the longer-term effect may be relatively positive as was the case with the 1755 Great Lisbon Earthquake and the 1812 Caracas, Venezuela Earthquake.

However, most natural disasters are not only catastrophic for the impacted area at the time, but they may leave scars that last for centuries. The decades to centuries of efforts in disaster reduction technology allow humans to reduce the impact of moderately strong events, evacuate people from the brunt of the disaster, and to greatly reduce the number of casualties. These reductions, however, are primarily only in industrialized countries and only in frequently impacted areas where preparations have been made. In all other areas, natural disasters are as deadly or worse than they have ever been, and they could change history. With the global human population having doubled since 1974, increasing by 4 billion,[1] there are many more people at risk of natural disasters than ever in history.

Some of these forthcoming natural disasters are similar or more impactful than the historical events. In areas where these events are common, through changes in technology, they are unlikely to change history. Natural disasters of the same magnitude that have occurred in the past in areas that are not ready for them could change history. Strong earthquakes in old cities with antiquities but no experience with earthquakes could be catastrophic and change history. If strong events strike areas in conflict, they will undoubtedly play a role in the outcome, thereby changing history. Finally, there is a chance that areas and the world could be struck by such catastrophic natural disasters that they change history up to a global scale. The human race has thrived, in part because we have not faced such events in historical times.

Earthquakes

Earthquakes will continue on a regular basis for as long as plate tectonics is operational. The frequency of earthquakes is a function of their magnitude; the higher the magnitude, the less frequent the earthquake. Earthquakes of magnitude 5 or less occur on a daily basis on the plate boundaries around the earth. The strongest earthquakes have magnitudes in the 9 range, and they are very rare.[2] However, strong earthquakes or "megaquakes" as they are called will continue in the future, and some may be powerful enough to change history depending on timing and location.

The strongest earthquake ever recorded is the Great Chilean or Valdivia earthquake. It occurred off the coast of Chile on May 22, 1960.[3] The magnitude was 9.5–9.6, and the duration was more than three minutes. The rupture was 21 miles (33 km) deep and located just offshore of Valdivia in the Pacific Ocean. The rupture occurred on the subduction zone surface as a megathrust earthquake. The ruptured area experienced up to 110 feet (30 m) of movement. The rapid uplift of the seafloor by faulting generated a large teletsunami that crossed the entire Pacific Ocean.

The deaths from the earthquake and tsunami in Chile were 5,700 to 7,000 with more than 2 million people left homeless. The tsunami caused more than $23.5 million in damage in Hawaii and caused two deaths and four injuries.

There was damage of $1 million on the west coast of the United States.[4] Four years later, the 1964 Good Friday earthquake struck Alaska. The magnitude of this earthquake was 9.2, making it the second most powerful earthquake in history and the most powerful in the northern hemisphere. It too caused a massive teletsunami.[5] Despite these two being most powerful earthquakes ever, they only caused the damage expected from much smaller events. The reason is they were in locations where less people would be impacted.

The next magnitude 9 earthquake was forty years later when the 2004 Indonesia earthquake and Indian Ocean tsunami struck and changed history. The 2011 Tohuku earthquake was also a magnitude 9 earthquake, and it too caused a major tsunami but did not change history because Japan is well-prepared for even strong earthquakes. Both of these were also megathrust earthquakes.[6]

All of these recent magnitude 9 earthquakes occurred on major subduction zones. Some researchers believed that only very large subduction zones could produce such large earthquakes but that might not be true. An "orphan" tsunami struck Japan on January 27 and 28, 1700, and it was described as exceptionally large waves. It was named an orphan because there were no local earthquakes, which could have triggered it. Historical reports claim that the waves ranged from 3.3 to 17.5 feet (1–5.4 m) and killed several people, destroyed buildings, flooded rice paddies, and damaged salt kilns. The source of the tsunami remained a mystery until 1996 when Japanese researchers studied the Cascadia Subduction Zone in Washington as the source. Calculating that the tsunami took about ten hours to travel to Japan, the megathrust earthquake occurred at about 9:00 p.m. on January 26, 1700.

American researchers found evidence of the earthquake and tsunami from Washington to Northern California. There are oral traditions among indigenous peoples from British Columbia to Northern California of shaking and flooding around that time. Archaeological studies found that several coastal villages were flooded and abandoned around 1700. The outermost tree rings of red cedar trees in Oregon and Washington show that they were killed by the sinking of coastal forests into the tidal zone. Carbon-14 dating of these trees shows that they died in the winter of 1699–1700. The coasts of northern California, Oregon, and Washington sunk by 3.3–6.6 feet (1–2 m). This sinking flooded the areas with seawater. Sedimentary layers on the old forests show sedimentation patterns consistent with earthquake shaking and flooding from a tsunami. Sediment cores from the nearby seafloor and samples from earthquake-induced landslides also support the timing of the event.[7]

The subsidence of the coastal forests indicates a large event. Generating a tsunami that crossed the entire Pacific Ocean and striking coastal Japan as high as 17.7 feet (5.4 m) also supports a large event. The large landslide scarps and widespread submarine deposits further support a large earthquake. The indigenous communities residing on the mountainside overlooking Pacheina Bay in southernmost British Columbia, Canada, were washed away up to approximately 75 feet (23 m) above sea level in about 1700. This is interpreted as the result of a tsunami reaching that height. Based on this evidence, researchers interpreted this earthquake to have occurred on the Cascadia Subduction Zone megathrust at an estimated magnitude of 8.7–9.2. It was named the Cascadia earthquake.[8]

The Cascadia earthquake was generated by the small Juan de Fuca Plate in the Pacific Ocean subducting under North America from Vancouver Island to Northern California. To produce an earthquake of magnitude 9, it has been calculated that the fault rupture must have been about 682 miles (1100 km) long and 62 miles (100 km) wide. It is postulated that severe shaking lasted about five minutes and severe aftershocks lasted for many months.

The problem now faced by the residents from northern California to Vancouver is that stress has been building up for more than 325 years without release in a major earthquake. The recurrence interval for magnitude 8+ earthquakes on this zone is calculated at about 300 years, which means they are overdue. The recurrence of magnitude 9 earthquakes are calculated at about

500 years. Based on this research, there is a 37 percent chance that the area will be struck by a magnitude 8.2+ earthquake within the next 50 years. There is a 10–15 percent chance that the entire Cascadia megathrust will rupture and generate a magnitude 9+ earthquake within 50 years.

The danger is that a large human population lives in this region including cities. Maps of evacuation routes were drawn based on a Cascadia tsunami striking the region. The study estimated that the at-risk population for areas within 0.6 miles (1 km) of the ocean exceeded 150,000 year-round residents, and it was done in 2000. These predictions are that the upcoming Cascadia earthquake and tsunami will impact millions of people, property, and infrastructure. One model predicts that casualties will exceed 10,000 with at least 30,000 injured. Economic losses will exceed $70 billion. It could be the worst natural disaster in US history. If it strikes, will it change global or US history? Probably not but it will change local history.

About 99 percent of earthquakes occur along plate boundaries, which for the continental United States is the West Coast with the Cascadia Subduction Zone in the north and the San Andreas Transform/fault in the south. In contrast, intraplate seismicity includes earthquakes that are not associated with plate boundaries. These earthquakes are impossible to predict with any certainty but are primarily low magnitude and cause no damage. This is fortunate because, unlike earthquake-prone areas, the areas they strike are completely unprepared for earthquakes. In the rare instances that even moderately large earthquakes strike areas that are unprepared, they can be devastating and change history.[9]

The best example of devastating intraplate earthquakes in the United States occurred in 1811 and 1812 in New Madrid, Missouri. A group of powerful earthquakes caused great damage. At the time, the Midwest was populated with a few settlers and Native Americans. As such, even though the New Madrid earthquakes were powerful, they only caused a moderate amount of damage and no change in history. However, the area of very strong shaking was an enormous 231,661 square miles (600,000 km^2). The area of moderate shaking was more than 1 million square miles (2.5 million km^2), and shaking was perceptible over more than 2 million square miles (5,000,000 km^2).[10]

The first of the major earthquakes struck on December 16, 1811, at 2:15 a.m. in northeast Arkansas. The magnitude was estimated at 7.5. The shaking woke people in New York City, Washington, D.C., and Charleston, South Carolina. The area that later became the city of Memphis, Tennessee, experienced extremely strong shaking. It was reported that sand and water was shot tens of feet into the air in fountains from the agitation. The town of Little Missouri was heavily damaged by the shaking. The earthquake generated a large seiche wave on the Mississippi River that bounced back and forth across the river. However, the sparse population limited damage to few man-made structures in the region. The second earthquake could be considered as a strong aftershock. It occurred the next morning on December 16 at 7:15 a.m. in northeast Arkansas. The magnitude has been estimated at 7.0, and it was felt all across the East Coast as well.[11]

The second main earthquake struck on January 23, 1812, at 9:15 a.m. in New Madrid, Missouri. The magnitude is estimated at 7.3. The damage in the epicenter included energetic water and sand fountains, extensive ground fissures, widespread landslides, caving of stream banks, and general ground warping. This earthquake is interpreted to have been caused by a rupture on the New Madrid North Fault. This movement on this fault may have increased the stress on the nearby Reelfoot Fault.

The final earthquake struck on February 7, 1812, at 3:45 a.m. also in New Madrid. The magnitude is estimated at 7.5, and it was generated by slip on the Reelfoot Fault of Missouri and Tennessee. The town of New Madrid was destroyed. Even as far away as St. Louis, Missouri, many houses were badly damaged or destroyed, and chimneys were toppled. Rapid uplift along part of the fault formed waterfalls on the Mississippi River at Kentucky Bend. Waves moved upstream, making it look like the river was flowing backward as reported by observers. The uplift also produced the Reelfoot Lake by blocking streams in Lake County, Tennessee.

These were not the only four earthquakes during this time. Between December 16, 1811, and March 15, 1812, there were more than 200 moderate to large aftershocks recorded by residents in the region. Seven to ten of the aftershocks are estimated to have had magnitudes greater than 6.0, and there were at least 100 aftershocks with magnitudes between 5.0 and 5.9.

There were many reports of surface deformation because of the shaking. These included widespread landslides on steep hillsides, areas of surface uplift, and even larger areas of subsidence that were commonly flooded. Large waves on the Mississippi River sunk several boats and washed others onto the shore and onto bluffs. Many riverbanks were undermined and several collapsed, and sandbars and parts of islands were washed away. In some cases, whole islands disappeared. The area most impacted by these surface failures was about 30,000–50,000 square miles (78,000–129,000 km^2) and extended from the current cities of Cairo, Illinois, to Memphis, Tennessee, and areas from Crowley's Ridge, Arkansas, to Chickasaw Bluffs, Tennessee. Evidence of the uplift and subsidence still exists today including Reelfoot Lake where the subsidence was 5 to 20 feet (1.5–6 m). Lake St. Francis in Arkansas, which is 40-mile (64-km) long by 0.6-mile (1-km) wide, was also formed by subsidence during prehistoric and the 1811–1812 earthquakes.[12]

Large sand blows formed by the eruption of a water and sand mixture at the surface from the 1811–12 earthquakes are also still evident. They formed by earthquake-driven liquefaction and are still found over an area of 4,015 square miles (10,400 km^2). The indications of liquefaction extend 124 miles (200 km) northeast of the New Madrid Seismic Zone, 155 miles (250 km) south to the mouth of the Arkansas River, and 150 miles (240 km) north–northwest nearly to St. Louis, Missouri.

Nothing close to the power of the four devastating earthquakes that occurred so close together has occurred in the New Madrid area since 1811–12, more than 212 years ago. Because these are intraplate earthquakes, it is impossible to predict when the next earthquake will strike with any certainty much less an earthquake of specific magnitude. This does not mean that this inevitable disaster has been forgotten. In November 2008, the US Federal Emergency Management Agency (FEMA) estimated that if an earthquake occurs in the New Madrid Seismic Zone like the 1811–12 earthquakes, it could produce the highest casualties and economic loss from a natural disaster ever in the history of the United States. The reports predicted catastrophic damage and loss of life across Alabama, Arkansas, Illinois, Indiana, Kentucky, Mississippi, Missouri, and particularly Tennessee. They predicted that an earthquake of magnitude 7.7 would damage or destroy tens of thousands of buildings, disrupt water and power distribution, cripple transportation systems, and other vital infrastructure. Predictions of death tolls are in the 250,000 range. The US Geological Survey estimates that the risk of an earthquake with a magnitude like the 1811–12 New Madrid earthquakes over the next fifty years is a very high 7–10 percent. The risk of a still devastating magnitude 6.0 earthquake over the next fifty years is a frightening 25–40 percent. These are extremely high percentages for earthquake predictions. The likelihood that an earthquake of this type will change history is high.[13]

Tsunamis

The 1960 and 1964 earthquakes and resulting tsunami disasters convinced the countries around the Pacific Ocean to develop the international Pacific Tsunami Warning System (PTWS). The margins of the Pacific Ocean are the most seismically and volcanically active areas on earth. This makes the Pacific Ocean the most prone to large tsunamis of any ocean basin. However, the advanced tsunami-sensing technology installed throughout the Pacific Ocean in addition to coordination centers, satellite warning systems, evacuation systems, and tsunami education programs of the PTWS greatly reduce the likelihood of tsunami disasters.[14]

PTWS incorporates technology from the Deep-Ocean Assessment and Reporting of Tsunamis (DART) program. DART is an early warning system that detects tsunamis while they are still far from the coast. The equipment consists of seafloor seismic sensors that detect tsunamis as they pass over. The sensors detect changes in the weight of water above them, which indicates that a tsunami is passing over. This information is transmitted to a buoy on the surface above it. The buoy transmits the information to a Geostationary Operational Environmental Satellite (GEOS), which then broadcasts tsunami warnings to land-based receivers all around the Pacific Ocean. The United States has its main receivers at the Pacific Tsunami Warning Center in Hawaii, the West Coast Tsunami Warning Center in California, and the Alaska Tsunami Warning Center. Coupled with information from the global seismic network, DART warns coastal communities about incoming tsunamis, which allows them to evacuate coastal areas in advance of the wave making landfall. Earthquake alerts are automatically issued to the tsunami warning centers for all earthquakes with magnitudes of 6.5 or more. Evacuation routes are well-marked and known by residents in all coastal communities designated by PTWS.[15]

Although the 1960 and 1964 earthquakes had the highest magnitudes ever recorded, the tsunamis they produced, although large, were not the biggest ever. That title is held by the 1958 Lituya Bay, Alaska Megatsunami. This monstrous wave was the result of a very unusual combination of an earthquake and a landslide. Its size was so large that it has since provided the backdrop for documentaries and movies on doomsday scenarios.

A major earthquake struck the Lituya Bay area in Alaska, about 120 miles (200 km) west of Juneau on July 9, 1958, at 10:16 p.m. The magnitude of the quake was measured at a strong 7.8. The rupture was 22 miles (35 km) deep on the Fairweather fault along the northeastern edge of Lituya Bay. This area is so desolate that normally, this earthquake would not have even been noticed by more than a few people, at most. However, the severe shaking loosened a large mass of rock high above the northeastern wall of the bay. The rock mass had a volume of 40 million cubic yards (30.6 million m^3) and a weight of 90 million tons (82 million mT). It dropped at least 3,000 feet (914 m) into the Gilbert Inlet of Lituya Bay as a rock fall and avalanche. It struck the water at more than 100 miles per hour (161 kph). The sound of this impact could be heard at least 50 miles (80 km) in all directions, and it caused about 1,312 feet (400 m) of ice to break off the Lituya glacier. The glacier released a large amount of sediment that slid into the bay, forming a double landslide. The combined masses caused water in the bay to slosh in a megatsunami or megaseiche that was the largest wave ever measured.[16]

The wave itself was 100 feet (30 m) high but traveled across the bay at such force that it continued up the facing slope to a phenomenal elevation of 1,720 feet (524 m). The wave is therefore not actually 1,720 feet (524 m) high as it is often credited. The climbing wave stripped away the trees and soil right down to the bedrock along the entire impacted bay wall. The removed trees and other material were washed into the bay and covered it with debris and ice. A few fishing boats were in the bay at the time. Some of them survived but others were destroyed, which resulted in a death toll of five. There is a story of a boat riding the crest of the wave up the hillslope, and it was deposited in the trees. The wave sloshed back and forth several times.

Megatsunami is not a scientific term but a Hollywood-type description of a very large tsunami. Disaster documentaries use the 1,720 feet (524 m) Lituya Bay wave as evidence for megatsunamis even though the wave was not that big. There is no evidence that a tsunami anywhere near that big has ever occurred in a large ocean. They appear to be restricted to small water bodies. A large asteroid impact in the ocean could produce a megatsunami, but large impacts are extremely rare. A very large volcanic explosion in the ocean could generate a huge tsunami, but likely not as big as the tsunami in Lituya Bay.[17]

There is a proposed future megatsunami in the Atlantic Ocean generated by the Cumbre Vieja volcano in the Canary Islands. There have been no recent tsunamis in the Atlantic Ocean, but there were recorded tsunamis in 1927 in eastern Canada and New England, in Puerto Rico in 1918, and a teletsunami from the 1755 Lisbon earthquake that reached all the way to the Caribbean.

The Cumbre Vieja volcano is on the island of La Palma which is part of the Canary Islands. The volcanoes on these islands are the result of a mantle plume, and they generally only produce dangerous eruptions on very rare occasions. The Canary Islands are off the west coast of Africa so quite a distance from the east coast of the United States. These many reasons make it unlikely that a megatsunami will be generated.

The only reason that a megatsunami is predicted is that during the 1949 eruption of the Cumbre Vieja volcano, the western flank of the mountain developed a slow-moving landslide. During the 1971 eruption of the volcano, the landslide slid farther downslope to an even more precarious position. A team of researchers from the United Kingdom studied the situation and speculated that the flank of the volcano could collapse during a subsequent eruption, producing a catastrophic avalanche into the Atlantic Ocean. This avalanche could produce a megatsunami similar to the 1958 megatsunami in Lituya Bay, Alaska. The research group further speculates that this megatsunami will be 300 feet (91 m) high and will travel across the Atlantic Ocean and devastate the east coast of the United States. The wave would also wreak havoc on the Caribbean islands and even west coast of Africa.[18] However, the Canary Islands have been erupting for many millions of years, and there is no evidence on the Atlantic coast or the Caribbean area ever having been struck by a major tsunami, much less a megatsunami.[19]

Hurricanes

There will always be hurricanes, typhoons, and cyclones, and they will always be destructive. The most devastating cyclone in history was Bhola in 1970, and it killed 300,000–500,000 people primarily in Bangladesh and changed history. However, there was another truly catastrophic cyclone that also struck Bangladesh in 1991, which took the lives of 138,866 people, and more recently, in 2008, Cyclone Nargis had a similar death toll of 138,373 in Myanmar. Therefore, truly horrific cyclones that are capable of changing history are still happening. The question is whether they are more or less likely.

Because hurricanes are strengthened if the ocean surface temperature is 79 °F (25 °C) and above, as the global temperatures and therefore surface temperatures rise, larger ocean areas are becoming capable of generating and supporting these large storms and for longer time during the year. This means that there should be more hurricanes. The overall trends support a general increase in these storms, but the number varies so much from year to year that it is not that obvious.[20]

However, considering typhoons that struck the Philippines alone, a trend becomes obvious. The country of the Philippines is comprised of a group of islands that sit in the primary Pacific Ocean typhoon track. It contains a huge amount of coastal area relative to the size of the country. For these reasons, it is both prone to typhoons and catastrophically impacted by them. The number of typhoons striking the Philippines has been increasing dramatically in recent history demonstrating a clear trend. It was struck by eighteen major typhoons in the 1950s and twenty-one in the 1960s but jumped to forty-seven in the 1970s, sixty-seven in the 1980s, forty-nine in the 1990s, and sixty-three in the 2000s. In the 2010s, however, it jumped to ninety-one. Similarly, the number of people killed by typhoons in the Philippines increased from 2,362 in the 1950s to 12,376 in the 2010s.[21] Part of the escalating death toll is because the population is increasing. However, the storms are also increasing in strength as exemplified by the Super Typhoon Haiyan or Yolanda in 2013, which was one of the most powerful storms ever.

Super Typhoon Haiyan began on November 2, 2013, as an area of low pressure several hundred miles east–southeast of Pohnpei, Micronesia. The low pressure moved west and it strengthened. By November 4, it had become a tropical storm and was named Haiyan. It underwent rapid intensification to typhoon intensity by November 5, and by the next day, it was a super typhoon at the equivalent strength of a category 5 on the Saffir–Simpson hurricane scale. By November 7, the maximum ten-minute sustained winds from Haiyan were measured at 145 miles per hour (230 kph) but estimated at 180 miles per hour (285 kph). The maximum one-minute sustained winds were measured at 195 miles per hour (315 kph), which made it the strongest typhoon ever, at least at the time. At 5:40 a.m. the next morning, Haiyan made its first landfall in Guiuan, Philippines at peak intensity. It then made five other landfalls in the Philippines before entering the South China Sea on November 9, where it turned and weakened.[22]

Super Typhoon Haiyan is also known as Super Typhoon Yolanda within the Philippines. It resulted in a death toll of 6,300 people, 1,061 people were missing, and another 28,689 people were injured. These casualties make it the second most devastating typhoon in the history of the Philippines. The ferocious winds caused a great amount of damage, but the storm surge at first landfall was the most devastating part.

Haiyan tracked westward across Leyte Gulf to Tolosa and then swung north into the funnel-shaped Bay of San Pedro. The funnel-shaped coastlines focused the storm surge. The storm surge was 10–14 feet (3–4 m) along the Leyte Gulf, but at the north end of Bay of San Pedro, tide gauges recorded a drop in sea level of about 3.3 feet (1 m) at 5:30 a.m. Seconds later, the storm surge sped through the bay and into the cities at tsunami speeds. The storm surge waves were 16 feet (5 m) high along the shores of the bay but 23–26 feet (7–8 m) in the city of Tacloban, reaching a peak of 46 feet (14 m) in the city of Guiuan.[23] Because the waves were not a result of an earthquake, it was not a true tsunami. Tsunami-like waves produced by storms are called meteotsunamis or meteorological tsunamis. This intense storm surge caused most of the deaths and devastation. The storm was made infamous by a documentary made by the Discovery Channel entitled “Megastorm: World’s Biggest Typhoon.”

The rapid intensification experienced by Super Typhoon Haiyan is now common. This is an unfortunate development. Category 1 hurricanes were formally just monitored with minor preparation in the landfall area. However, category 1 hurricanes can now intensify to category 5 in just twenty-four hours.[24] This means all hurricanes must be considered potentially catastrophic regardless of their current status. This means that evacuations and other very expensive and disruptive preparations must be made for all storms. The reason for this change in the hurricane, typhoon, and cyclone development is the increase in surface temperature of the ocean with climate change.[25] The higher the surface temperature, the quicker the storm can intensify with no other mitigating factors. This is especially dangerous on ocean islands where evacuation may be impossible. This scenario has the potential to produce a natural disaster that could change history in poorly prepared areas but is unlikely in prepared areas.

One of the recent catastrophic results of early and rapid intensification of a hurricane was Hurricane Maria in 2017. Hurricane Maria began as a tropical wave that formed off the western coast of Africa on September 12, 2017. It moved westward across the Atlantic Ocean, slowly organizing into a storm. It developed into a tropical depression on September 16, about 665 miles (1,070 km) east of Barbados. The surface temperature of the ocean was a warm 84 °F (29 °C), which encouraged the rapid intensification to a tropical storm just six hours later. By late on September 17, Maria became a category 1 hurricane. Then Maria underwent among the most rapid intensifications recorded, increasing its wind speed from 85 to 165 miles per hour (140–270 kph) in just twenty-four hours. This increase was from category 1 to 5 in one day. At this point, it was located 15 miles (25 km) east–southeast of Dominica on September 18. Category 5 hurricane Maria made landfall on the island of Dominica on September 19. This was the first time in history that Domenica was struck by such a strong hurricane. Interaction with land caused Hurricane Maria to weaken to category 4, but it restrengthened to category 5 when it entered the Caribbean Sea. At this point, the sustained wind speed was 175 miles per hour (280 kph), which was the peak wind speed for Maria.[26]

At this point, Maria underwent eyewall replacement, which weakened it to category 4 just before it made landfall near Yabucoa, Puerto Rico. This landfall was on September 20, 2017, at 6:15 a.m. with a sustained wind speed of 155 miles per hour (250 kph), making it the most powerful hurricane to strike Puerto Rico since 1928. The ocean island was completely devastated by Maria, which weakened as it crossed. Once it was over the warm Atlantic Ocean during the afternoon, it began to restrengthen to a major hurricane. It reached a second peak strength on September 22 when the wind speed reached 125 miles per hour (205 kph) to the north of Hispaniola.

When Hurricane Maria underwent eyewall replacement, the eye increased in size from 10 to 32 miles (16–51.5 km) in diameter. As a result, the storm's eyewall, the most dangerous part of the storm, covered the entire island of Puerto Rico. Heavy rain reached nearly 38 inches (96.5 cm) in one day in one area, causing serious flooding. Flood waters rose more than 6 feet (1.8 m) in just 30 minutes in one area, reaching a depth of at least 15 feet (4.6 m). These rains produced landslides on hillslopes all across the island at as much as twenty-five landslides per square mile (10 landslides/km^2). Along the coast, storm surge caused a maximum inundation of 6–9 feet (1.8–2.7 m).

Forecasts warned that Hurricane Maria would likely strike the island, and evacuations were ordered in advance of the storm. This resulted in a 20 percent increase in travel out of Puerto Rico the day before it made landfall. Government officials opened 450 storm shelters during the afternoon of September 18, and at least 2,000 people took shelter in them by September 19.

The storm destroyed the entire power grid in Puerto Rico, leaving 100 percent of the 3.4 million residents without power. The storm knocked out 95 percent of telephone cell networks, and 85 percent of the above-ground phone and internet cables were destroyed. The weather radar antenna of Puerto Rico was destroyed, and it was designed to withstand winds of 130 miles per hour (209 kph). It took more than nine months to replace it. Greater than 80 percent of the island's agriculture was destroyed including the whole coffee industry.

The recovery from the impact of Hurricane Maria was very slow. As of September 26, 95 percent of Puerto Rico was still without electricity, more than half the population was still without tap water, and 95 percent did not have cell phone service. By October 6, 89 percent of homes had no electricity, 44 percent did not have tap water, and 58 percent of the residents had no cell phone service. By October 20, even a month after the storm, about 3 million people or 88 percent of the residents were still without electricity, about 1 million people or roughly 29 percent still had no tap water, and 40 percent were without cell service. By November 17, nearly two months after the hurricane, only 392 miles (631 km) of the island's 5,073 miles (8,164 km) of road were passable, only half the sewage plants were operating, and at least 60,000 houses still were without roofs. Three months after the storm, at least 1.5 million people or 45 percent of the population were still without electricity, and 14 percent had no tap water. Four months after the storm, 450,000 customers were still without electricity, and regular blackouts increased that number. Many residents abandoned Puerto Rico as a result. By November 5, at least 100,000 people had abandoned Puerto Rico and were residing in the mainland United States. By December 17, that number was up to 130,000.[27]

Maria was the deadliest hurricane in Puerto Rico since 1899 and the deadliest in Dominica since 1834. The total death toll of the disaster was 3,059 people with 2,975 killed in Puerto Rico alone. This makes Hurricane Maria the second-worst natural disaster in the United States in terms of the number of deaths. The total cost of the damage made it the third costliest hurricane in US history.

Therefore, a recent storm with all of the technology to detect and track hurricanes, organized evacuation routes and shelters, as well as construction methods to reduce damage was still no match for a hurricane that underwent rapid intensification. Similar havoc was wreaked on North Carolina in 2024 by Hurricane Helene, which also underwent rapid intensification.[28] Considering that hurricanes are now commonly intensifying faster than action can be taken to prepare for them, a hurricane, typhoon, or cyclone natural disaster impactful enough to change history is becoming more probable in unprepared areas.

Tornadoes

The main producer of tornadoes is the United States with an average of about 1,200 tornadoes per year.[29] The reason for the large number of tornadoes is the violent meeting of Gulf of Mexico warm, moist air with cold, dry air from Canada. Most tornadoes were in Tornado Alley from Texas northward across the Central Plains. However, with time, the main area of tornado production has expanded to the north and east and developed in the southeast, which now produces as many and as strong tornadoes as Tornado Alley. Therefore, the area threatened by deadly tornadoes has grown larger, which increases the possibility of a history-changing tornado. The strongest tornado ever recorded was in Moore, Oklahoma, in 1999, but whether they are becoming more powerful with time is unclear, as none have been as powerful since then. However, this tornado convinced many people with federal assistance to install safe rooms and tornado shelters in their homes. This will reduce the number of casualties in a tornado, reducing the likelihood of a history-changing tornado natural disaster significantly. Another aspect of danger from tornadoes results from tornado super outbreaks. This is when many tornadoes are generated in a single weather situation. The 1974 Super Outbreak certainly changed history, but another powerful Super Outbreak also occurred in 2011.[30]

In terms of sheer number of tornadoes, the 2011 Super Outbreak was, by far, the largest tornado outbreak in US history. From April 25 to 28, 2011, an extensive weather system produced 360 confirmed tornadoes. On April 27 alone, 216 tornadoes were spawned, the largest single-day total ever. Tornadoes struck the South, Midwest, and Northeast including twenty-one states from Texas to New York and even southern Canada. Even the advances in forecasting, warning systems, and public preparedness in response to previous outbreaks were no match for this massive event. This tornado outbreak was deadly and the costliest in history.[31]

The outbreak started with an upper-level low-pressure trough in the Southern Plains on April 25 and it strengthened on April 26. Discrete tornado-bearing supercells developed by the afternoon and early evening with damaging winds and hail overnight. On April 27, the storms intensified and moved east into the Ohio, Mississippi, and Tennessee River valleys. It finally drifted east into the Atlantic Ocean on April 28. The NOAA Storm Prediction Center issued fifty-six major severe weather watches over the four days including forty-one tornado watches, and there were fifteen severe thunderstorm watches. Of the tornado watches, ten were designated as "particularly dangerous situations."[32]

One "particularly dangerous situation" tornado watch was issued at 3:25 p.m. on April 25 for most of Arkansas and parts of Missouri, Oklahoma, Texas, and Louisiana. Several EF2 tornadoes and one EF3 tornado struck Arkansas causing destruction, but most tornadoes were minor. In all, forty-two tornadoes struck that day, causing five deaths and fifty-nine injuries. April 26 was similarly unstable but with two primary areas of activity. Tornado watches were issued for the Great Lakes region by the afternoon. Supercells in central Michigan produced strong thunderstorms and at least two tornadoes. As the system moved east, severe thunderstorms in Pennsylvania and New York produced 2-inch hail and one EF1 tornado. A southern center of activity produced stronger supercells with numerous tornadoes across Texas, Louisiana, and Arkansas. Several tornadoes were at EF2 strength and even EF3 strength in Kentucky, but most were weak. There were a total of fifty-five tornadoes on April 26.

April 27 was, by far, the most catastrophic day of the 2011 super outbreak. There was essentially continuous activity between April 26 and 27 with nonstop severe thunderstorms and tornadoes from the evening of April 26 through the late morning of April 27. A line of storms moved across Louisiana, Mississippi, Alabama, and Middle and East Tennessee early in the day, but they strengthened eastward into Alabama. Most of these tornadoes were weak, especially earlier in the day, but there were several deadly EF2, EF3, and one EF4 tornadoes in eastern Mississippi and Alabama.

However, a second powerful line of supercells developed in the late morning and early afternoon. It moved into Mississippi and produced a deadly EF5 tornado near Philadelphia, Mississippi. This tornado was so powerful that it stripped pavement off roads and removed the top 2 feet (0.61 m) of soil in places. It was one of four EF5 tornadoes produced by this line. An EF4 tornado developed in this group in eastern Mississippi and crossed into Alabama. At 1:45 p.m., a "particularly dangerous situation" tornado watch was issued for most of Alabama and parts of Mississippi, Tennessee, and Georgia. This produced a number of powerful tornadoes, especially in Alabama. Many were deadly including a multivortex tornado in Cullman, Alabama. The most famous tornado of the 2011 Super Outbreak was the Tuscaloosa to Birmingham, Alabama EF4 tornado. This tornado touched down at 5:10 p.m. and remained on the ground for at least ninety minutes. It carved an 80-mile (129 km) long swath of destruction. The huge tornado was seen on several television stations in the area, and it was rebroadcasted throughout the United States. This tornado was responsible for killing sixty-five people, which was the most of any single storm.

Powerful tornadoes continued east into Tennessee through the evening. A separate line of severe storms developed and extended from central and northern Virginia north through Maryland, Pennsylvania, and New York in the afternoon and evening. There were also tornado watches in southern Canada. Tornadoes developed all along the East Coast from North Carolina to

New York into early April 28. Additional tornadoes also occurred from Florida through Pennsylvania and southern New York during April 28. The entire weather system finally drifted into the Atlantic Ocean after this.[33]

The 2011 Super Outbreak set numerous records as the greatest tornado outbreak ever. There were 360 tornadoes over a four-day period with 216 on April 27 alone. This was the highest number of tornadoes in a single day ever. Of the 360 tornadoes, 238 were in Alabama alone. The six National Weather Service offices most impacted by the outbreak issued 303 tornado warnings on April 27, and the Alabama office issued ninety of them. Tornadoes tracked for a total of 3,200 path miles (5,150 km) on the ground for the entire outbreak. One EF5 tornado was continuously on the ground for 132 miles (212 km). The death toll was 348 people across six states including 324 deaths from tornadoes directly, which made this the deadliest tornado outbreak in at least a decade. It also caused more than 3,100 injuries and a total of $12 billion in damage, making it the costliest in US history.

Considering the that United States experiences an average of 1,200 tornadoes per year, having 360 in four days is exceptional. This outbreak was relatively recent, so it could be argued that tornado outbreaks are becoming larger. Certainly, the area susceptible to regular tornado strikes is growing larger with time. It can therefore be argued that tornadoes are becoming more dangerous and that a history-changing outbreak is possible. On the other hand, because the number of safe rooms is increasing, even if there are even more tornadoes, they may cause less casualties. The cost, however, will continue to increase.

Volcanoes

It is unlikely that small to moderately large volcanic eruptions will change history. Precursors are well studied, and active dangerous volcanoes are well monitored. Eruptions are generally well anticipated, and evacuations can be carried out effectively. The eruption can cause significant damage to buildings and infrastructure at costs of hundreds of thousands of dollars. Large volcanic eruptions that cause climate change, on the other hand, can change history. Explosive eruptions that inject large amounts of aerosols into the upper troposphere and stratosphere will significantly cool the climate for up to a few years. Such eruptions have occurred in recent history with catastrophic effects. With the burgeoning global population, eruptions like the 1783 Laki eruption in Iceland or the 1812 Tambora eruption that produced the year without a summer could have history-changing consequences.

However, these eruptions are small compared with those from supervolcanoes. The term "supervolcano" was coined in a British television program on powerful volcanoes rather than being a true scientific term. It is generally defined as a volcano that erupts in an exceptional explosion and the ejection of immense quantities of aerosols and pyroclastic material. Supervolcano eruptions can change the Earth's atmosphere for an extended period of time. Humans have been lucky that a real supervolcano has not erupted during modern history or civilization would likely have collapsed, and the human species may have even gone extinct.[34]

The most recent supervolcano eruption was from the Toba volcano in Indonesia. Toba is on Sumatra in a chain of volcanoes. The last eruption occurred 74,000 years ago as previously described. This eruption was an 8 out of 8 on the volcanic explosivity index, producing about 700 million cubic miles (2,800 km^3) of ejecta. As a comparison, the 1980 Mount St. Helens violent eruption yielded about 0.24 cubic miles (1 km^3) of ejecta, and it was exceptionally destructive to that part of Washington. The Toba eruption is interpreted to have caused a major bottleneck in the human race through its impact on the worldwide climate. There is also a supervolcano in the United States in the Yellowstone area.

The beautiful geysers and thermal springs of Yellowstone National Park in Wyoming hide a violent volcanic past with the potential of reawakening with catastrophic results. The hydrothermal activity in the park is generated by a large reservoir of magma at depth. This igneous activity results from North America moving over a mantle plume or hot spot similar to Hawaii at about 1.1 inches (2.7 cm) per year. There is a record of igneous activity progressing from the Columbia River plateau that began about 16.5 million years ago across the Snake River Plain to the location of the current activity. These volcanic centers track the hot spot over time. There is evidence of supervolcano eruptions in Idaho prior to them reaching the Yellowstone area. The Bruneau-Jarbidge volcanic caldera formed about 10–12 million years ago and erupted depositing ash at least one foot (30 cm) deep at least 1,000 miles (1,600 km) away in Nebraska where it killed herds of rhinoceros and camels, among others, which are now preserved in Ashfall Fossil Beds State Historical Park.[35]

Yellowstone experienced small rhyolitic eruptions early in its volcanic activity about 8.7 million years ago. This minor volcanic area evolved into a major caldera with three rhyolite supervolcano eruptions. The eruptions produced ash, pumice, and rock fragments that formed extensive sheets as thick as 1,312 feet (400 m) in some areas. The first of the three eruptions occurred about 2.08 million years ago. The large developing magma chamber under the area suffered a catastrophic collapse of the roof of the chamber. This created the huge Big Bend Ridge, Snake River, and Red Mountains calderas that are 121–153 miles by 64–97 miles (75–95 km by 40–60 km) as segments of the huge Island Park caldera. This large caldera covered an area of about 1,000 square miles (2,560 km^2). The collapse generated a mammoth eruption that shot at least 610 cubic miles (2,500 km^3) of ejecta into the air. The huge ash cloud spread across the United States and covered 5,985 square miles (15,500 km^2), forming the Huckleberry Ridge Tuff deposit.[36]

The second Yellowstone eruption was about 1.3 million years ago and was the smallest. This eruption produced the 10 to 17-mile (16–27.2 km)-wide Henry's Fork caldera. The eruption shot about 70 cubic miles (292 km^3) of ejecta into the air, and it covered an area of 1,042 square miles (2,700 km^2)

and now called the Mesa Falls Tuff. This is too small to be a supervolcano. On the other hand, the third and final eruption occurred about 630,000 years ago. This eruption produced the large Yellowstone caldera, which is about 30 miles wide by 53 miles long (48 km by 85 km). This supervolcano erupted and expelled 240 cubic miles (1,000 km^3) of ejecta that covered an area of 2,896 square miles (7,500 km^2). This deposit is the Lava Creek Tuff with a maximum thickness of 590–660 feet (180–200 m).

The concern is whether Yellowstone will erupt in the near future. As a result, there has been considerable research done on the caldera. Yellowstone ranks number twenty-first on the greatest volcanic threats in the United States with a threat score at the top of the high threat category. Studies show that the main magma chamber under Yellowstone is 50 miles (80 km) long by 24 miles (40 km) wide, and it extends from 3.1 to 9.9 miles (5–16 km) in depth. This enormous volume of magma is available for eruption. There is also a lower crustal reservoir of mafic that extends from 12 miles to more than 25 miles (20–40 km) deep. It is about 4.5 times the volume of the upper magma reservoir. It generates enough heat to keep the upper reservoir full of magma and available for a major eruption. The floor of the Yellowstone caldera rose almost 3 inches (7.6 cm) per year between 2004 and 2008. This rate was more than three times the fastest rate observed since 1923 when measurements began.

Impacts

Of all the natural disasters that could change history, asteroid impacts with the earth can do the most damage in the shortest amount of time. Fortunately, they are very rare. The point of impact can be pulverized and ignited by the extreme heat. If they impact the ocean, they can generate larger tsunamis than any earthquake or volcano. An impact can raise enough dust to blot out the sun, causing significant global cooling. The most recent impact was the 2013 Chelyabinsk, Siberia event, and although it was spectacular, it was small and did only superficial damage. However, a little more than a century earlier, there was a much more impressive impact.

In 1908, the Tunguska Asteroid impact struck a remote area of Siberia. Unfortunately, the area was so sparsely populated at the time that there were few observations. It was also so long ago that the level of scientific observations was not advanced enough at the time to conduct the necessary level of research. This event was also apparently not a true surface impact but instead an atmospheric explosion. For these reasons, most of the information on size, speed, and altitude had to be calculated through modeling. Because the area is so sparsely populated and there were no human casualties, this is not a natural disaster for humans nor could it change history, though much wildlife perished.[37]

On June 30, 1908, at 7:17 a.m., a column of blue light as brilliant as the sun was seen crossing the sky. At 7:27 a.m., a blinding flash was followed by a loud explosion. There were shock waves emerging from the explosion that broke windows and knocked people down for hundreds of miles in all directions. The blast flattened the forest in a huge area of 830 square miles (2,150 km^2), with the shape of a butterfly having a wingspan of 43 miles (70 km) and a body length of 34 miles (55 km). It is estimated that at least 80 million trees were snapped like toothpicks. The air blast from the explosion was registered on barometers between ground zero all the way to England. Dense clouds of dust were raised to high altitudes above the area, and they reflected incoming light. As a consequence, the night skies glowed brightly in Europe and Asia for quite a while after the event. There were reports of the glow being so bright that some people in Asia could read newspapers outdoors by the glow until midnight.

Modeling was later done using the observations and conditions of the blast zone to quantify the mechanics of the event. It appears that the explosion was caused by the air burst of a stony meteoroid since no crater or debris was ever found. The meteoroid was approximately 160–620 feet (50–190 m) in diameter and weighed 220 million pounds (100 million kg), making it quite substantial. The asteroid entered the Earth's atmosphere at a velocity of at least 34,000 miles per hour (55,000 kph). The heat generated by air resistance was intense that it caused the asteroid to heat to more than 44,500 °F (24,710 °C) and ultimately to explode in midair. It is estimated that the explosion took place at an altitude of between 4 and 6 miles (6–10 km) to as much as 15 miles (24 km).

The estimated power of the explosion was equivalent to between 10 and 15 megatons of TNT (42–63 PpJ) and as much as 30 megatons of TNT (130 PpJ). The blast from a 30-megaton (130-PpJ) explosion is approximately equal to the energy from detonation of about 2,000 Hiroshima-size atomic bombs. The shock wave from the explosion would have registered as a magnitude 5.0 earthquake on seismographs had they been in use. An explosion of this magnitude would have the ability to destroy a large city, leveling all the buildings and infrastructure. Fortunately, this event was in a remote area and has not happened since. Estimates suggest that Tunguska-type events occur once in about 1,000 years.

Another model that also matches the observations and results hypothesizes that a large iron-rich asteroid glanced off the earth's atmosphere and returned into a solar orbit. The explosion would have been the glancing blow of the asteroid in this case rather than its destruction. This solution would explain why no asteroid debris was ever found in the explosion area.

If the Tunguska explosion had occurred over a populated area, it likely would have caused a natural disaster that changed history. However, Tunguska was miniscule compared to other impacts that have occurred directly on earth rather than an airburst explosion. These impacts can do more than change history; they can and have changed the composition of life on the planet. Fortunately, they are very rare. The most famous of these impacts is the Chicxulub impact in the Gulf of Mexico.

Previously, one of the big questions in geology was why the dinosaurs went extinct. Walter Alvarez set out to answer this question by traveling around the world and sampling and analyzing the sedimentary rocks that crossed the time when the dinosaurs went extinct. He found high amounts of iridium in these rocks, and on that basis, he proposed that a large asteroid collided with the earth at that time. This impact is proposed to have caused the dinosaurs to go extinct.

The question then became if such a large asteroid collided with the earth, why don't we see the impact crater? The answer to this question is because the impact crater is hidden on the seafloor. The crater was later identified in the southern Gulf of Mexico near the Yucatán Peninsula. Using geophysical techniques, researchers located the large 93–112 mile (150–180 km) diameter and 12 mile (20 km) deep crater buried under thick sediments that were deposited on top. The crater was previously drilled by several oil companies, but they interpreted the melted rocks and breccias as having been produced through volcanic activity. However, samples taken from a borehole at depths of 0.75 and 0.8 miles (1.2 and 1.3 km) into the imaged crater were re-examined and contained compelling evidence of an impact. Another well found 820 feet (250 m) of breccia (crushed rock) in the crater overlying 1,247 feet (380 m) of melt rock, both confirming an impact origin for the structure.[38]

The crater was named Chicxulub, the Mayan word for "tail of the devil." The crater is buried underneath 0.6 miles (1 km) of sediments, which is why it took so long to find and explore. The crust beneath the crater is crushed and fractured to a depth of about 6.2–18.6 miles (10–30 km). The age of the impact was isotopically dated at 66 million years old, which is the Cretaceous–Tertiary (K-T) boundary.[38]

The collision generated sedimentary and impact deposits all around the Gulf of Mexico and even beyond. There is a 17-inch (0.5 m) thick deposit in Haiti that is composed primarily of melt spherules and shocked quartz. Both of these indicate that they were formed in an impact crater. Shocked quartz has been found as far away as New Jersey. The Arroyo el Mimbral in Mexico contains several yards (meters) in thickness that has a lower part composed of altered melt spherules, and it is overlain by sand with abundant plant debris. The seafloor in this location was 1,625 feet (500 m) deep in the ocean at the time. Plants and their debris were not growing at that depth, so they had to be swept there by large waves. The top of this deposit contains anomalous levels of iridium, which is also a result of the impact. There are thinner deposits in coastal Texas and Alabama that reflect the impact of large waves.[39]

The characteristics of the Chicxulub impact were modeled based on the size, shape, and features of the Chicxulub crater, the impact deposits around the Gulf of Mexico, and comparison with other documented impacts. The Chicxulub asteroid was a fragment of a larger asteroid of about 106 miles (170 km) across. The asteroid fragment that produced the Chicxulub crater was between 7 and 50 miles (11–81 km) across, and it had a mass of between 1.1×10^{12} and 5.1×10^{14} tons (1.0×10^{15} - 4.6×10^{17} kg). There is evidence that other asteroid fragments struck in a row to the northeast across the Atlantic Ocean. The velocity of impact is estimated at 44,640 miles per hour (72,000 kph). The explosion of impact delivered an estimated energy equivalent to between 21 and 921 billion Hiroshima atomic bombs or about 100 million megatons of TNT.[40]

The explosion of the impact generated shock waves with energy equivalent to a magnitude-10 earthquake. The temperature in the core of this explosion is estimated to have been more than 18,000 °F (10,000 °C). The impact is estimated to have produced an air blast that radiated in all directions with winds of at least 621 miles per hour (1,000 kph). The blast and winds scoured soils, shredded vegetation, and killed every animal in a blast zone between 560 and 1,120 miles (900–1,800 km) across. Debris strewn by the impact is estimated to have rained down all across the Gulf of Mexico and the entire Caribbean region. Depending upon the distance from the impact, this debris ranged from rocky rubble to impact-melted spherules. It is estimated that this ejecta was at least 1,000 feet (305 m) thick around the impact site, and it thinned with distance in all directions.

The impact generated tsunamis that shot in all directions across the Gulf of Mexico, the developing Caribbean Sea, and the Atlantic Ocean. They devastated the nearby coastlines. The estimated wave heights range from 164 to 984 feet (50–300 m). They crashed onto the shorelines across the Gulf of Mexico and penetrated inland as much as an estimated 62 miles (100 km) depending on location. Once the waves flooded inland, they swept massive amounts of continental debris back into the sea. The incoming waves and backwash deeply eroded the seafloor, and the shaking caused landslides all across the seafloor.

The impact also produced extensive wildfires, dust clouds to great height, acid rain, aerosols, destruction of the ozone layer, and huge quantities of greenhouse gases. The acid rain was nitric acid from the heating of the atmosphere and sulfuric acid from the asteroid and from vaporizing sulfur-bearing minerals at the impact. The damage caused by acids on rocks, and life is evident in the sedimentary deposits. Excessive soot caused by the burning of vegetation is also abundant in these deposits. The dust and aerosols from the impact and extensive soot from wildfires choked the atmosphere blotting out the sun. The fossils show that photosynthetic plankton experienced a global extinction at that time. It is hypothesized that sunlight could not reach the surface through this thick cloud, which shut down photosynthesis across the globe. This shutdown of photosynthesis also occurred in land plants. As a result, the herbivores that fed on these plants, and, in turn, the carnivores that preyed on the herbivores all experienced a major extinction event.[41]

The atmosphere cooled several degrees immediately after the impact because of the heavy suspended particulates and aerosols. This cooling also contributed to the extinctions. However, once the dust settled and the skies cleared, the temperature increased. The cause of the rising temperatures was the increase in greenhouse gases resulting from the vaporization of materials at the impact and the resulting wildfires. The estimates of the warming range from 2.7 °F (1.5 °C) to an astonishing 13.5 °F (7.5 °C). This increase would have caused a major disruption in the remaining ecosystems.

Impacts are thankfully very rare because if a large asteroid like that which produced Chixulub strikes the earth, it will at the very least change history. It could wipe out the human race. Even a relatively small body like that which caused Tunguska could change history if it strikes the right location. Because they are so rare, asteroid impacts are not worth worrying about. If an asteroid on a collision course with the earth is identified early enough, there might be the ability to change its course using weapons. But this is more of a scenario for movies.

The Takeaway

There are some natural disasters that will change history in any time or location. These include asteroid impacts and supervolcanoes. Fortunately, neither of these has occurred during the past 10,000 years or so in recorded history. The 1908 Tunguska impact could have caused a natural disaster if it had occurred in a populated location, and, in that case, it would have certainly changed history. Otherwise, the last time there was an earth-changing impact was many millions of years ago. Therefore, although impacts are very devastating, they are so rare that despite the many disaster movies about them, they are extremely unlikely. Similarly, supervolcano eruptions are also very rare. Toba erupted 74,000 years ago and Yellowstone, 630,000 years ago. These two were the last to erupt, so again, regardless of the disaster movies, they are extremely unlikely.

More common major volcanic eruptions like Laki or those in 536 CE can change the earth's climate for several years and cause major upheavals to society. These can cause changes to history if they are large enough. Fortunately, global trade can smooth over the changes caused by such eruptions for several years. But that is the best society could weather at this point. It would take a volcanic eruption or eruptions that disrupt the climate for a longer period like those around 536 CE to definitely change history.

Other than these rare events, preparation for more normal natural disasters is possible, but in many cases, it costs a lot to be effective. Most active volcanoes in populated areas are carefully monitored, and evacuations are carried out well before the eruptions either eliminating casualties or reducing them by 100–1,000-fold. Similarly, tsunami early warning systems are now worldwide and are coupled with education and evacuation routes. Unless the tsunami is generated close to shore, they give coastal residents ample warning to evacuate, and casualties can be greatly reduced or eliminated. Earthquake hazard reduction engineering in California and Japan is outstanding and reduces the casualties from earthquakes by 100–1,000-fold compared to less prepared or unprepared areas, which, unfortunately, are far more common. Tornado safe rooms reduce casualties significantly if people can get to them before the tornado strikes. Global trade can redistribute food during droughts, thereby reducing their impact.

Can natural disasters still change history? Absolutely, yes. However, the impact of many natural disasters that could have changed history has been greatly reduced in many areas. With improved natural hazard reduction engineering and technology and more widespread adoption of these advances, natural disasters can become less impactful and be even less likely to change history.

NOTES

Chapter 1

1 National Severe Storm Laboratory, *Severe Weather 101: Tornado Basics.* [cited 3/20/2025]; Available from: https://www.nssl.noaa.gov/education/svrwx101/tornadoes/.

2 Gates, A.E., *Earth's Fury: Science of Natural Disasters* (New York: Wiley & Sons-Blackwell Publishing, 2022), 403.

3 Lichtenberg, F., *Why Has Longevity Increased More in Some States than in Others? The Role of Medical Innovation and Other Factors* (New York: Manhattan Institute, 2007).

4 Gates, *Earth's Fury*, 403.

5 Dietrich, J., "Coping with Disasters in Antiquity and the Bible: Practical and Mental Strategies," in *Past Vulnerability: Volcanic Eruptions and Human Vulnerability in Traditional Societies Past and Present*, ed. I.F. Riede (Aarhus Universitetsforlag: The Netherlands, 2015), 151–67.

6 Dilek, Y. and Kahya, O., "Flood and Earthquake as Punishment of Gods in Antiquity," *Journal of Disaster and Risk* 6, no. 3 (2023): 819–28.

7 Gates, *Earth's Fury*, 403.

8 Ritchie, H., L. Rodés-Guirao, E. Mathieu, M. Gerber, E. Ortiz-Ospina, J. Hasell, and M. Roser, *Population Growth* (2023) [cited 2/15/2025]; Available from: https://ourworldindata.org/population-growth.

9 Barberi, F., Mathew S. Davis, Roberto Isaia, Rosella Nave, and Tullio Ricci, "Volcanic Risk Perception in the Vesuvius Population," *Journal of Volcanology and Geothermal Research* 172, no. 3–4 (2008): 244–58.

10 Roth, D., *Louisiana Hurricane History* (US: National Weather Service, 2010), 70.

11 Daniell, J. E., B. Khazai, and F. Wenzel, "Uncovering the 2010 Haiti Earthquake Death Toll," *Natural Hazards and Earth System Sciences Discussions* 1, no. 3 (2013): 1913–42.

12 Gates, A. E., and D. Ritchie, *E-book: Encyclopedia of Earthquakes and Volcanoes*, Fourth Edition (New York: Facts On File, Inc., 2016), 368.

13 Blake, E.S. and Gibney, E.J., *The Deadliest, Costliest, and Most Intense United States Tropical Cyclones from 1851 to 2010 (and Other Frequently Requested Hurricane Facts)* (Miami and Asheville, NC: National Hurricane Center, 2011).

14 Zoraster, R.M., “Vulnerable Populations: Hurricane Katrina As a Case Study,” *Prehospital and Disaster Medicine* 25, no. 1 (2010): 74–8.

15 Gates and Ritchie, *E-book*, 368.

16 Ambraseys, N., *Earthquakes in the Mediterranean and Middle East: A Multidisciplinary Study of Seismicity up to 1900* (London: Cambridge University Press, 2009).

17 Pararas-Carayannis, G., “The Great Tangshan Earthquake of 28 July 1976 in China—Analysis of Tsunami Generation in the Bohai and Yellow Seas,” *Journal of Tsunami Society International* 40, no. 2 (2021): 122–45.

18 Evans, S. G., Nicholas J. Roberts, Anatoli Ischuk, Keith B. Delaney, Galina S. Morozova, and Olga Tutubalina, “Landslides Triggered by the 1949 Khait Earthquake, Tajikistan, and Associated Loss of Life,” *Engineering Geology* 109, no. 3–4 (2009): 195–212.

19 Sidorin, A.Y., “On the 70th Anniversary of the 1949 Khait Earthquake in Tajikistan,” *Seismic Instruments* 56 (2020): 491–500.

20 Yablokov, A., “The Tragedy of Khait: A Natural Disaster in Tajikistan,” *Mountain Research and Development* 21, no. 1 (2001): 91–3.

21 Pruce, J.R., “Benefit Concerts, Constituencies of Compassion, and the Culture Industry,” *The Mass Appeal of Human Rights* (2019): 39–76.

22 DesRoches, R., Mary Comerio, Marc Eberhard, Walter Mooney, and Glenn J. Rix, “Overview of the 2010 Haiti Earthquake,” *Earthquake Spectra* 27, no. 1 (2011): 1–21.

23 Calais, E., Andrew Freed, Glen Mattioli, Falk Amelung, Sigurjón Jónsson, Pamela Jansma, Sang-Hoon Hong, Timothy Dixon, Claude Prépetit, and Roberte Momplaisir, “Transpressional Rupture of an Unmapped Dault during the 2010 Haiti Earthquake,” *Nature Geoscience* 3, no. 11 (2010): 794–9.

24 Gates and Ritchie, *E-book*, 368.

25 Margesson, R., and Maureen Taft-Morales, *Haiti Earthquake: Crisis and Response* (Washington, DC: Congressional Research Service, 2010).

26 Daniell, Khazai, and Wenzel, “Uncovering the 2010 Haiti Earthquake Death Toll.”

27 Bilham, R., “The 1737 Calcutta Earthquake and Cyclone Evaluated,” *Bulletin of the Seismological Society of America* 84, no. 5 (1994): 1650–7.

28 Chakrabarti, R., “The Calcutta Cyclone of 1737: Cyclone or Earthquake?” in *Force of Nature* (Routledge, 2017), 123–36.

29 Zschau, J., and Andreas N. Küppers, eds., *Early Warning Systems for Natural Disaster Reduction* (New York: Science & Business Media, 2013).

30 Morrison, J., “Cities Around the Globe Are Eagerly Importing a Dutch Speciality—Flood Prevention,” *Smithsonian Magazine*, 2019, https://www.smithsonianmag.com/

innovation/cities-around-globe-eagerly-importing-dutch-speciality-flood-prevention-180973679/.

31 Gates, *Earth's Fury*, 403.

32 Earthquake Hazards Program, "M 6.9 October 17, 1989 Loma Prieta Earthquake," 2024 [cited 2024 11/19]; Available from: https://www.usgs.gov/programs/earthquake-hazards/science/m-69-october-17-1989-loma-prieta-earthquake.

33 Aagaard, Brad T., G.C.Beroza, "The 1906 San Francisco Earthquake a Century Later: Introduction to the Special Section," *Bulletin of the Seismological Society of America* 98, no. 2 (2008): 817–22.

Chapter 2

1 Boyd, O. S., William D. Barnhart, James Bourke, Martin Chapman, Paul S. Earle, Guo-chin Dino Huang, Jessica A. Thompson Jobe, Won-Young Kim, Frederik Link, Mairi Litherland, Andrew Lloyd, Maureen D. Long, Sara McBride, Andrew J. Michael, Walter D. Mooney, Gregory S. Mountain, Sissy Nikolaou, Alexandros Savvaidis, Felix Waldhauser, Cecily J. Wolfe, and Clara Yoon, *Preliminary Observations of the 5 April 2024 4.8 New Jersey Earthquake* (USGS: Reston, VA, 2024), 240–50.

2 Voiland, A., *Tracking Canada's Extreme 2023 Fire Season*, Earth Observatory 2023 [cited 2/12/2025]; Available from: https://earthobservatory.nasa.gov/images/151985/tracking-canadas-extreme-2023-fire-season.

3 Gates, A.E. and D. Ritchie, *E-book: Encyclopedia of Earthquakes and Volcanoes*, Fourth Edition (New York: Facts On File, Inc., 2016), 368.

4 Normand, J.C.L., and E. Heggy, "Assessing Flash Flood Erosion Following Storm Daniel in Libya," *Nature Communications* 15 (2024): 6493.

5 Chowdhury, S., "Saudi Arabian Desert's First Ever Snowfall Shown in Photos Published," *Newsweek*, 2024.

6 Burt, J.E. and E. Aguado, *Understanding Weather & Climate, Global Edition*, Seventh Edition (New York: Pearson, 2015), 608.

7 Choi, C.Q., "History's Greatest El Niño May Have Caused Severe 19th Century Famine," *Inside Science*, 2017.

8 Burt and Aguado, *Understanding Weather & Climate*, 608.

9 National Severe Storms Laboratory, "Severe Weather 101: Tornado Forecasting" [cited 3/16/2025]; Available from: https://www.nssl.noaa.gov/education/svrwx101/tornadoes/forecasting/.

10 Gates, A.E., *Earth's Fury: Science of Natural Disasters* (Hoboken, NJ: Wiley & Sons-Blackwell Publishing, 2022), 403.

11 Burt and Aguado, *Understanding Weather & Climate*, 608.

12 Balaguru, K., Gregory R. Foltz, and L. Ruby Leung, "Increasing Magnitude of Hurricane Rapid Intensification in the Central and Eastern Tropical Atlantic," *Geophysical Research Letters* 45, no. 9 (2018): 4238–47.

13 Hellin, J., Martin Haigh, and Frank Marks, "Rainfall Characteristics of Hurricane Mitch," *Nature* 399, no. 6734 (1999): 316–19.

14 Singer, M., *2013 State of the Climate: Record-Breaking Super Typhoon Haiyan*, 2014.

15 Lynch, A., "Case Study: Banqiao Dam (China, 1975)," *Damfailures.org*, 2016, https://damfailures.org/case-study/banqiao-dam-china-1975/.

16 WMO, "World's Deadliest Tropical Cyclone Was 50 Years Ago," *World Meteorological Organization News*, 2020.

17 Fritz, H.M., Christopher D. Blount, Swe Thwin, Moe Kyaw Thu, and Nyein Chan, "Cyclone Nargis storm surge in Myanmar," *Nature Geoscience* 2, no. 7 (2009): 448–9.

18 Gates, *Earth's Fury*, 403.

19 Erickson, J., *Plate Tectonics: Unraveling the Mysteries of the Earth* (New York: Facts on File Inc., 2014), 305.

20 Gates, and Ritchie, *E-book*, 368.

21 Gates, *Earth's Fury*, 403.

22 Ibid.

23 Searle, M.P. and Treloar, P.J., "Introduction to Himalayan Tectonics: A Modern Synthesis," *Geological Society, London, Special Publications* 483 (2019): 1–17.

24 Nemčok, M., S. Rybár, S.T. Sinha, S.A. Hermeston, and L. Ledvényiová, "Transform Margins: Development, Controls and Petroleum Systems—An Introduction," *Geological Society, London, Special Publications* 431 (2016): 1–38.

25 Gates, *Earth's Fury*, 403.

Chapter 3

1 Gates, A.E., *Earth's Fury: Science of Natural Disasters* (Hoboken, NJ: Wiley & Sons-Blackwell Publishing, 2022), 403.

2 Schulte, P., Laia Alegret, Ignacio Arenillas, José A. Arz, Penny J. Barton, Paul R. Bown, Timothy J. Bralower, et al., "The Chicxulub Asteroid Impact and Mass Eextinction at the Cretaceous-Paleogene Boundary," *Science* 327, no. 5970 (2010): 1214–18.

3 Cui, Y. and L.R. Kump, "Global Warming and the End-Permian Extinction Event: Proxy and Modeling Perspectives," *Earth-Science Reviews* 149 (2015): 5–22.

4 Gates, A.E., and D. Ritchie, *E-book: Encyclopedia of Earthquakes and Volcanoes* (New York: Facts on File, Inc., 2016), 368.

5 Breining, G., *Super Volcano: The Ticking Time Bomb Beneath Yellowstone National Park* (Voyageur Press, 2007).

6 Rampino, M.R. and S. Self, "Bottleneck in Human Evolution and the Toba Eruption," *Science* 262, no. 5142 (1993): 1955–5.

7 Gathorne-Hardy, F., and W. Harcourt-Smith, "The Super-Eruption of Toba: Did It Cause a Human Bottleneck?" *Journal of Human Evolution* 43, no. 3 (2003): 227–30.

8 Gates and Ritchie, *E-book*, 368.

9 Friedrich, W.L., et al., *Fire in the Sea: The Santorini Volcano: Natural History and the Legend of Atlantis* (Cambridge: Cambridge University Press, 2000).

10 Freewalt, J., "The Theran Disruption: The Minoan Eruption of Thera and Its Possible Impact on Civilisations," *World History Seminar–HIST510 A001*, Spring 2013.

11 Şahoğlu, V., et al., "Volcanic Ash, Victims, and Tsunami Debris from the Late Bronze Age Thera Eruption Discovered at Çeşme-Bağlararası (Turkey)," *Proceedings of the National Academy of Sciences* 119, no. 1 (2022): e2114213118.

12 Foster, K.P., R.K. Ritner, and B.R. Foster, "Texts, Storms, and the Thera Eruption," *Journal of Near Eastern Studies* 55, no. 1 (1996): 1–14.

13 McCoy, F.W. and G. Heiken, "Tsunami Generated by the Late Bronze Age Eruption of Thera (Santorini), Greece," *Pure and Applied Geophysics* 157, no. 6 (2000): 1227–56.

14 Pearson, C.L., et al., "Annual Radiocarbon Record Indicates 16th century BCE Date for the Thera Eruption," *Science Advances* 4, no. 8 (2018): eaar8241.

15 Freewalt, "The Theran Disruption."

16 Booysen, R., *Thera and the Exodus: The Exodus Explained in Terms of Natural Phenomena and the Human Response to It* (Winchester: John Hunt Publishing, 2013).

17 Gates and Ritchie, *E-book*, 368.

18 Booysen, *Thera and the Exodus*.

19 Ali, R.U., "Medieval Europe: The Myth of Dark Ages and the Impact of Islam," *Islamic Studies* 51, no. 2 (2012): 155–68.

20 Gibbons, A., "Why 536 Was 'The Worst Year to Be Alive,'" *Science* 362, no. 6416 (2018): 733–4.

21 Arjava, A., "The Mystery Cloud of 536 CE in the Mediterranean Sources," *Dumbarton Oaks Papers* 59 (2005): 73–94.

22 Büntgen, U., et al., "Cooling and Societal Change During the Late Antique Little Ice Age from 536 to around 660 AD," *Nature Geoscience* 9, no. 3 (2016): 231–6.

23 Helama, S., et al., "Volcanic Dust Veils from Sixth Century Tree-Ring Isotopes Linked to Reduced Irradiance, Primary Production and Human Health," *Scientific Reports* 8, no. 1 (2018): 1339.

24 Toohey, M., et al., "Climatic and Societal Impacts of a Volcanic Double Event at the Dawn of the Middle Ages," *Climatic Change* 136 (2016): 401–12.

25 Newfield, T.P., "The Climate Downturn of 536–50," *The Palgrave Handbook of Climate History* (2018): 447–93. doi:10.1057/978-1-137-43020-5_32

26 Breeze, A., "Arthur's Battles and the Volcanic Winter of 536–37," *Northern History* 52, no. 2 (2016): 161–72.

27 Helama, et al., "Volcanic Dust Veils from Sixth Century Tree-Ring Isotopes," 1339.

28 Ganiev, R.T., "The Climate Cooling of 536–545 AD and the Rise of the Turkic Empire," *Life Science Journal* 11, no. 12 (2014): 496–9.

29 Toohey, et al., "Climatic and Societal Impacts of a Volcanic Double Event," 401–12.

30 Thordarson, T., and S. Self, "The Laki (Skaftár Fires) and Grímsvötn Eruptions in 1783–1785," *Bulletin of Volcanology* 55 (1993): 233–63.

31 Thordarson, T., and S. Self, "Atmospheric and Environmental Effects of the 1783–1784 Laki Eruption: A Review and Reassessment," *Journal of Geophysical Research: Atmospheres* 108, no. D1 (2003): AAC 7-1–AAC 7-29.

32 Jackson, E.L., "The Laki Eruption of 1783: Impacts on Population and Settlement in Iceland," *Geography* 67, no. 1 (1982): 42–50.

33 Sigurdsson, H., "Volcanic Pollution and Climate: The 1783 Laki Eruption," *Eos, Transactions American Geophysical Union* 63, no. 32 (1982): 601–2.

34 Schmidt, A., et al., "Excess Mortality in Europe Following a Future Laki-Style Icelandic Eruption," *Proceedings of the National Academy of Sciences* 108, no. 38 (2011): 15710–15.

35 Sigurdsson, "Volcanic Pollution and Climate."

36 Schmidt, et al., "Excess Mortality in Europe."

37 Gates, *Earth's Fury*, 403.

38 Global Volcanism Program, "Report on Rabaul (Papua New Guinea)," in *Bulletin of the Global Volcanism Network*, ed. E. Venzke (Washington DC: Smithsonian Institution, 1994).

39 Blong, R., "The Rabaul Eruption, 1994," *The Australian Geographer* 25, no. 2 (1994): 186–90.

40 USGS Communications and Publishing, *Remembering Mount Pinatubo 25 Years Ago*, 2016.

41 Newhall, C.G., *The Cataclysmic 1991 Eruption of Mount Pinatubo, Philippines* (Reston, VA: USGS, 1997).

42 International Society of Decade Volcanoes, "Research at Decade Volcanos Aimed at Disaster Prevention," *Eos, Transactions American Geophysical Union* 75, no. 30 (1994): 340–50.

43 Meara, R., et al., "Magmatic Memories: Eldfell, 1973," *Volcanica* 7, no. 1 (2024): 361–403.

44 Williams Jr, R.S., and J.G. Moore, *Man against Volcano: The Eruption on Heimaey, Vestmann Islands, Iceland* (US Government Printing Office, 1976).

45 Williams Jr, R.S., *Lava-Cooling Operations during the 1973 Eruption of Eldfell Volcano, Heimaey, Vestmannaeyjar, Iceland* (US Geological Survey, 1997).

Chapter 4

1 Gates, A.E., *Earth's Fury: Science of Natural Disasters* (Hoboken, NJ: Wiley & Sons-Blackwell Publishing, 2022), 403.

2 Earthquake Hazards Program, *Earthquake Magnitude, Energy Release, and Shaking Intensity*, USGS, https://www.usgs.gov/programs/earthquake-hazards/earthquake-magnitude-energy-release-and-shaking-intensity.

3 Papanastassiou, D., "Seismic Hazard Assessment in the Area of Mystras-Sparta, South Peloponnesus, Greece, Based on Local Seismotectonic, Seismic, Geologic Information and on Different Models of Rupture Propagation," *Natural Hazards* 18 (1998): 237–51.

4 Pararas-Carayannis, G., "The Earthquake and Tsunami of July 21, 365 AD in the Eastern Mediterranean Sea: Review of Impact on the Ancient World: Assessment of Recurrence and Future Impact," *Science of Tsunami Hazards* 30, no. 4 (2011): 253–92.

5 Papanastassiou, D., K. Gaki-Papanastassiou, and H. Maroukian, "Recognition of Past Earthquakes along the Sparta Fault (Peloponnesus, Southern Greece) during the Holocene, by Combining Results of Different Dating Techniques," *Journal of Geodynamics* 40, nos. 2–3 (2005): 189–99.

6 Roberts, J.T., *The Plague of War: Athens, Sparta, and the Struggle for Ancient Greece* (New York: Oxford University Press, 2017), 448.

7 Liu, J., "The Social Structure of the Spartan City-State: A Game-theoretic Explanation," *Ancient History Bulletin* 36, no. 3–4 (2022): 179–92.

8 Papanastassiou, "Seismic Hazard Assessment."

9 Roberts, *The Plague of War*, 448.

10 Papazarkadas, N., S.B. Pomeroy, S.M. Burstein, W. Donlan, J.T. Roberts, "A Brief History of Ancient Greece: Politics, Society, and Culture," in *The Classical Review* (New York: Oxford University Press, 2004), 360.

11 Robinson, A., "Earthquakes in Political, Economic, and Cultural History," *Oxford Research Encyclopedia of Natural Hazard Science*, 2017.

12 Liu, "The Social Structure of the Spartan City-State."

13 Roberts, *The Plague of War*, 448.

14 deBoer, J., and D. Sanders, *Earthquakes in Human History: The Far-Reaching Effects of Seismic Disruptions* (Princeton, NJ: Princeton University Press, 2005), 100.

15 Martínez-Loriente, S., V. Sallarès, and E. Gràcia, "The Horseshoe Abyssal Plain Thrust Could be the Source of the 1755 Lisbon Earthquake and Tsunami," *Communications Earth & Environment* 2, no. 1 (2021): 145.

16 Fonseca, J.F.B.D., "A Reassessment of the Magnitude of the 1755 Lisbon Earthquake," *Bulletin of the Seismological Society of America* 110, no. 1 (2020): 1–17.

17 Oliveira, C.S., "Review of the 1755 Lisbon Earthquake Based on Recent Analyses of Historical Observations," *Historical Seismology: Interdisciplinary Studies of Past and Recent Earthquakes* (2008): 261–300.

18 Zitellini, N., et al., "Source of 1755 Lisbon Earthquake and Tsunami Investigated," *Eos, Transactions American Geophysical Union* 82, no. 26 (2001): 285–91.

19 Oliveira, "Review of the 1755 Lisbon Earthquake."

20 Silva, P., et al., "Earthquake Environmental Effects of the AD 1755 Lisbon-Earthquake-Tsunami in Spain," *Actas IX Reunión Cuaternario Ibérico* (Faro: 2017): 53–7.

21 Oliveira, "Review of the 1755 Lisbon Earthquake."

22 Dynes, R.R., "The Lisbon Earthquake in 1755: The First Modern Disaster," in *The Lisbon Earthquake of 1755: Representations and Reactions*, ed. T.E.D. Braun and J.B. Radnor (New York: Oxford University Press, 2005), 34–49.

23 Pereira, A.S., "The Opportunity of a Disaster: The Economic Impact of the 1755 Lisbon Earthquake," *The Journal of Economic History* 69, no. 2 (2009): 466–99.

24 Dynes, "The Lisbon Earthquake in 1755."

25 Sanides-Kohlrausch, C., "The Lisbon Earthquake, 1755: A Discourse about the 'Nature' of Nature," in *Is Nature Ever Evil?* (Routledge, 2003), 106–19.

26 Araújo, A.C., "The Lisbon Earthquake of 1755: Public Distress and Political Propaganda," *E-journal of Portuguese History* 4, no. 1 (2006): 1–11.

27 Anderson, F., *Crucible of War: The Seven Years' War and the Fate of Empire in British North America, 1754–1766* (Vintage, 2007).

28 Sanides-Kohlrausch, "The Lisbon Earthquake, 1755."

29 Hamacher, W., *Premises: Essays on Philosophy and Literature from Kant to Celan* (Cambridge, MA: Harvard University Press, 1996), 263.

30 Oliveira, "Review of the 1755 Lisbon Earthquake."

31 Robinson, A., *Earth-Shattering Events: Earthquakes, Nations, and Civilization* (New York: Thames & Hudson, 2016).

32 Butterworth, H., *South America: A Popular History of the Struggle for Liberty in the Andean Republics and Cuba* (New York: Doubleday and McClure, 1898).

33 Altez, R., "New Interpretations of the Social and Material Impacts of the 1812 Earthquake in Caracas, Venezuela," in *Ancient Earthquakes*, ed. M. Sintubin, I.S. Stewart, T.M. Niemi, and E. Altunel (Boulder, CO: Geological Society of America Special Paper, 2010), 47–58.

34 Lucca, R.A., "In 1812, a National Catastrophe Helped Topple a Weak Government," *Caracas Chronicles*, 2019.

35 McCook, S., "Nature, God and Nation in Revolutionary Venezuela: The Holy Thursday Earthquake of 1812," in *Aftershocks: Earthquakes and Popular Politics in Latin America*, ed. J.a.L.J. Buchenau (University of New Mexico Press: Albuquerque, NM, 2009), 43–69.

36 Butterworth, *South America.*

37 Robinson, *Earth-Shattering Events.*

38 McCook, *Nature, God and Nation in Revolutionary Venezuela.*

39 Gates, A.E. and D.Ritchie, *E-book: Encyclopedia of Earthquakes and Volcanoes* (New York: Facts on File, Inc., 2016), 368.

40 Nyst, M., T. Nishimura, F.F. Pollitz, and W. Thatcher, "The 1923 Kanto Earthquake Reevaluated Using a Newly Augmented Geodetic Data Set," *Journal of Geophysical Research: Solid Earth* 111, no. B11 (2006). doi:10.1029/2005JB003628.

41 Gates, *Earth's Fury*.

42 Takemura, M., "Two Big Aftershocks Immediately after the 1923 Kanto Earthquake Strong Ground Motions and Locations of the Epicenters," *Journal of Geography (Chigaku Zasshi)* 108, no. 4 (1999): 440–57.

43 Kaneko, H., "Evaluation of Tsunami Disasters Caused by the 1923 Great Kanto Earthquake," *Journal of Disaster Research* 18, no. 6 (2023): 578–89.

44 Hammer, J., "The Great Japan Earthquake of 1923," *Smithsonian Magazine*, 2011.

45 Endo, R., and Junko Iwahashi, "Analysis of the Spatial Distribution of the Landslides Triggered by the 1923 Great Kanto Earthquake, Japan," *Proceedings of the Japan Academy Series B* 100, no. 2 (2024): 123–39.

46 Schencking, J.C., "The Great Kanto Earthquake and the Culture of Catastrophe and Reconstruction in 1920s Japan," *The Journal of Japanese Studies* 34, no. 2 (2008): 295–31.

47 Scawthorn, C., Tomoaki Nishino, J. Charles Schencking, and Janet Borland, "Kantō Daikasai: The Great Kantō Fire Following the 1923 Earthquake," *Bulletin of the Seismological Society of America* 113, no. 5 (2023): 1902–23.

48 Borland, J., "Capitalising on Catastrophe: Reinvigorating the Japanese State with Moral Values through Education following the 1923 Great Kantô Earthquake," *Modern Asian Studies* 40, no. 4 (2006): 875–907.

49 Schencking, "The Great Kanto Earthquake."

50 Orihara, M., and Gregory Clancey, "The Nature of Emergency: The Great Kanto Earthquake and the Crisis of Reason in Late Imperial Japan," *Science in Context* 25, no. 1 (2012): 103–26.

51 Borland, "Capitalising on Catastrophe."

52 Seidensticker, E., *Low City, High City: Tokyo from Edo to the Earthquake* (New York: Knopf, 1983).

53 Schencking, "The Great Kanto Earthquake."

54 Orihara and Clancey, "The Nature of Emergency."

55 Schencking, "The Great Kanto Earthquake."

56 Orihara and Clancey, "The Nature of Emergency."

Chapter 5

1 Burt, J.E., and E. Aguado, *Understanding Weather & Climate, Global Edition*, Seventh Edition (New York: Pearson, 2015), 608.

2 Gates, A.E., *Earth's Fury: Science of Natural Disasters* (Hoboken, NJ: Wiley & Sons-Blackwell Publishing, 2022).

3 Dolin, E.J., "Did Hurricanes Save America?" *American Heritage* 65, no. 5 (2020), https://www.americanheritage.com/did-hurricanes-save-america.

4 Mulcahy, M., *Hurricanes and Society in the British Greater Caribbean, 1624–1783* (Baltimore, MD: John Hopkins University Press, 2005).

5 Ghosh, P., "1780: The Deadliest Atlantic Hurricane Season Ever," *International Business Times*, 2011.

6 Korten, T., "The Bahamas and the Caribbean Have Withstood Hurricanes for Centuries," *Smithsonian Magazine*, 2019, https://www.smithsonianmag.com/history/bahamas-and-caribbean-have-withstood-hurricanes-centuries-180973157/.

7 Alsop, E., "Great Hurricane of 1780," *ArcGIS Story Maps*, 2022, https://storymaps.arcgis.com/stories/c0d6275a6b5245b6a8de6fa378f2a8d9.

8 Neely, W., *The Great Hurricane of 1780: The Story of the Greatest and Deadliest . . .* (iUniverse, 2012), 312.

9 Coleman, J.S.M. and LaVoie, S.A., "09 Paleotempestology: Reconstructing Atlantic Tropical Cyclone Tracks in the Pre-HURDAT Era," in *Modern Climatology* (Salt Lake City, UT: Utah State University, 2012), 237–56.

10 Neely, *The Great Hurricane of 1780: The story of the Greatest and Deadliest . . .* 312.

11 Sledge, J., "Solano's Storm: Venture Inside the 18th-century Gulf Hurricane that Wrecked a Spanish Invasion Fleet," *Mobile Bay*, 2020.

12 Emanuel, K., *Divine Wind: The History and Science of Hurricanes* (New York: Oxford University Press, 2005).

13 Mulcahy, *Hurricanes and Society in the British Greater Caribbean.*

14 Dolin, "Did Hurricanes Save America?"

15 Ramos, M.G., "After the Great Storm: Galveston's Response to the Hurricane of 1900," *Texas Almanac*, 2008, https://www.texasalmanac.com/articles/galvestons-great-hurricane.

16 Van Riper, A.B., "The Great Galveston Hurricane of 1900," *The Ultimate History Project*, 2013.

17 Burnett, J., "The Tempest At Galveston: 'We Knew There Was a Storm Coming, but We Had No Idea," 2017, https://www.npr.org/2017/11/30/566950355/the-tempest-at-galveston-we-knew-there-was-a-storm-coming-but-we-had-no-idea.

18 McElreath, D.H., Daniel Adrian Doss, Carl Jensen, Hilliard Lackey, Don W. Jones, Mike Wigginton, and Rebecca Goza, "Dangers from the Sea: Considerations of the 1900 Galveston Hurricane," *International Journal of Maritime History* 29, no. 3 (2017): 529–43.

19 Dittrick, P., "Galveston Was 'The Ellis Island of the West'," *United Press International*, 1991.

20 Zalzal, K.S., "Benchmarks: September 8, 1900: Massive Hurricane Strikes Galveston, Texas," *Earth Magazine*, 2016.

21 NOAA, *1900 Major Hurricane UNNAMED (1900239N15318)*. IBTrACS [cited 3/23/2025], https://ncics.org/ibtracs/index.php?name=v04r01-1900239N15318.

22 Ramos, "After the Great Storm."

23 NOAA, "The Great Galveston Hurricane of 1900," in *NOAA Celebrates 200 Years of Science, Service and Stewardship* (Washington, DC: National Oceanic and Atmospheric Administration, 2007), https://celebrating200years.noaa.gov/magazine/galv_hurricane/welcome.html#intro.

24 Roker, A., *The Storm of the Century: Tragedy, Heroism, Survival, and the Epic True Story of America's Deadliest Natural Disaster: The Great Gulf Hurricane of 1900* (New York: William Morrow, 2015), 320.

25 Bixel, P.B., and E.H. Turner, *Galveston and the 1900 Storm* (Austin: University of Texas Press, 2000).

26 NOAA, "The Great Galveston Hurricane of 1900."

27 Burnett, "The Tempest at Galveston."

28 National Park Service, *Galveston Hurricane of 1900*, May 30, 2019 [cited 2025 3/23/2025], https://www.nps.gov/articles/galveston-hurricane-of-1900.htm.

29 Van Riper, *The Great Galveston hurricane of 1900*.

30 Ramos, "After the Great Storm"

31 Bixel and Turner, *Galveston and the 1900 Storm*.

32 NOAA, "45th Anniversary of the Bhola Cyclone," 2015 [cited 3/23/2025], https://www.aoml.noaa.gov/hurricane_blog/45th-anniversary-of-the-bhola-cyclone/.

33 Frank, N., and S. A. Husain, "The Deadliest Tropical Cyclone in History?" *Bulletin of the American Meteorological Society* 52, no. 6 (1971): 438–45.

34 Siddik, M.A., Md. Ashiqur Rahman, and Md. Abdul Majed Sajib, "The Human Cost of Decadal Cyclones in Bangladesh: A Comparative Study," *International Journal of Energy, Environment, and Economics* 30, no. 4 (2022): 427–45.

35 Faisal, Q.A., Md Arif Hossain, and S.M. Quamrul Hassan, "Simulation of Storm Surge and its Variation for Bhola Cyclone using MRI Storm Surge Model," *The Atmosphere* 9 (June 2022): 54–62.

36 NOAA. "45th Anniversary of the Bhola Cyclone."

37 Eskander, S.M., and Edward B. Barbier, "Long-Term Impacts of the 1970 Cyclone in Bangladesh," *World Development* 152, no. 4 (2022): 105793.

38 Sarkar, J., *Ploughshares and Swords: India's Nuclear Program in the Global Cold War* (Ithaca: Cornell University Press, 2022).

39 Gates, *Earth's Fury*.

40 Siddik, Rahman, and Sajib, "The Human Cost of Decadal Cyclones in Bangladesh."

Chapter 6

1 Burt, J.E., and E. Aguado, E., *Understanding Weather & Climate, Global Edition*, 7th ed. (New York: Pearson, 2015), 608.

2 Gates, A.E., *Earth's Fury: Science of Natural Disasters* (Hoboken, NJ: Wiley & Sons-Blackwell Publishing, 2022), 403.

3 National Weather Service, *1925 Tri-State Tornado*, 2025 [cited 3/23/2025], https://www.weather.gov/pah/1925Tornado.

4 Akin, W.E., *The Forgotten Storm: The Great Tri-State Tornado of 1925* (Guilford, CT: Lyons Press, 2002), 224.

5 Felknor, P., *The Tri-State Tornado: The Story of America's Greatest Tornado Disaster* (iUniverse, 2004), 140.

6 Maddox, R.A., M.S. Gilmore, C.A. Doswell III, R.H. Johns, C.A. Crisp, D.W. Burgess, J.A. Hart, and S.F. Piltz, "Meteorological Analyses of the Tri-State Tornado Eevent of March 1925," *E-Journal of Severe Storms Meteorology* 8, no. 1 (2013): 1–27.

7 Johns, R.H., D.W. Burgess, C.A. Doswell III, M.S. Gilmore, J.A. Hart, and S.F. Piltz, "The 1925 Tri-State Dornado Damage Path and Associated Storm System," *Electronic Journal of Severe Storms Meteorology* 8, no. 2 (2013): 1–33.

8 Grazulis, T., *Significant Tornadoes, 1880–1989: Volume 1, Discussion and Analysis* (St. Johnsbury, VT: Environmental Films, 1991).

9 Mason, A., *Death Rides the Sky: The Story of the 1925 Tri-State Tornado* (Rockford, IL: Black Oak Media, 2011).

10 Partlow, G., *America's Deadliest Twister: The Tri-State Tornado of 1925* (Carbondale, IL: Southern Illinois University Press, 2014).

11 National Weather Service, *1925 Tri-State Tornado*.

12 Mason, *Death Rides the Sky*.

13 National Weather Service, *1925 Tri-State Tornado.*

14 Johns, Burgess, Doswell, Gilmore, Hart, and Piltz, "The 1925 Tri-State Tornado Damage Path and Associated Storm System."

15 Changnon, S. A., "Two Record Long Tornadoes in Illinois and Their Major Impacts," *Transactions of the Illinois State Academy of Science* 102, no. 3–4 (2009): 175–80.

16 Powell, S., "A Beast from the Sky: The Tri-State Tornado of 1925," *Midwest Mayhem*, 2023.

17 Grice, G.K., R.J. Trapp, S.F. Corfidi, R. Davies-Jones, C.C. Buonanno, J.P. Craven, K.K. Droegemeier, et al., "The Golden Anniversary Celebration of the First Tornado Forecast," *Bulletin of the American Meteorological Society* 80, no. 7 (1999): 1341–8.

18 Grazulis, T.P., *Significant Tornadoes 1974–2022* (St. Johnsbury, VT: The Tornado Project, 2023).

19 Corfidi, S.F., Steven J. Weiss, John S. Kain, Sarah J. Corfidi, Robert M. Rabin, and Jason J. Levit, "Revisiting the 3–4 April 1974 Super Outbreak of Tornadoes," *Weather and Forecasting* 25, no. 2 (2010): 465–510.

20 Levine, M., *American Tornado: Devastation, Survival, and the Most Violent Tornado Outbreak of the Twentieth Century* (Ebury, 2007), 307.

21 Locatelli, J.D., Mark T. Stoelinga, and Peter V. Hobbs, "A New Look at the Super Outbreak of Tornadoes on 3–4 April 1974," *Monthly Weather Review* 130, no. 6 (2002): 1633–51.

22 Corfidi, Weiss, Kain, Corfidi, Rabin, and Levit, "Revisiting the 3–4 April 1974 Super Outbreak of Tornadoes."

23 Levine, *American Tornado.*

24 Grazulis, T.P., *Significant Tornadoes 1974–2022* (St. Johnsbury, VT: The Tornado Project, 2023).

25 National Center for Environmental Information, *April 3–4, 1974 Super Outbreak*, 2024 [cited 3/23/2025]; NOAA, April 3–4, 1974 Super Outbreak, https://www.ncei.noaa.gov/news/april-3-4-1974-super-outbreak.

26 Government of the United States, *Public Law 93-288–May 22, 1974* (Washington, DC: U. Congress, 1974).

27 Erdman, J., "8 Things the April 1974 Super Outbreak Taught Us About Tornadoes," *The Weather Channel*, April 3, 2020, https://weather.com/safety/tornado/news/2020-04-03-april-1974-super-outbreak-tornadoes.

28 National Weather Service, "The Great Plains Tornado Outbreak of May 3–4, 1999—Storm A Information," 1999, https://www.weather.gov/oun/events-19990503.

29 Risk Management Solutions Report, "The 1999 Oklahoma Tornado Outbreak: 10-Year Retrospective," *Risk Management Solutions Inc.*, 2009, 14.

30 Simmons, K.M., Paul Kovacs, and Gregory A. Kopp, "Tornado Damage Mitigation: Benefit–Cost Analysis of Enhanced Building Codes in Oklahoma," *Weather, Climate, and Society* 7, no. 2 (2015): 169–78.

31 Mathis, N., *Storm Warning: The Story of a Killer Tornado* (New York: Touchstone, 2007), 256.

32 Erdman, J., "One of America's Most Infamous Tornadoes, the Deadly Oklahoma City F5, Struck 25 Years Ago," *Tornado Central* 2024.

33 National Weather Service, "The Great Plains Tornado Outbreak of May 3–4, 1999."

34 Comstock, R.D., and Sue Mallonee, "Comparing Reactions to Two Severe Tornadoes in One Oklahoma Community," *Disasters* 29, no. 3 (2005): 277–87.

35 Merrell, D., Kevin M. Simmons, and Daniel Sutter, "Taking Shelter: Estimating the Safety Benefits of Tornado Safe Rooms," *Weather and Forecasting* 17, no. 3 (2002): 619–25.

36 OK Office of Emergency Management, "Safe Rooms in Oklahoma—Did You Know?" [cited 2025 3/24/2025], https://oklahoma.gov/oem/emergencies-and-disasters/2013/20130518-severe-weather-events/safe-rooms-in-oklahoma.html.

37 Standohar-Alfano, C.D., Seamus Freyne, Andrew J. Graettinger, Royce W. Floyd, and Thang N. Dao, "Performance of Residential Shelters during the May 20, 2013, Tornado in Moore, Oklahoma," *Journal of Performance of Constructed Facilities* 29, no. 5 (2015): 1229–37.

38 Doswell, C.A., and Harold E. Brooks, "Lessons Learned from the Damage Produced by the Tornadoes of 3 May 1999," *Weather and Forecasting* 17, no. 3 (2002): 611–18.

39 FEMA, "Safe Room Funding," 2024 February 13, 2025 [cited 2025 3/24/2025], https://www.fema.gov/emergency-managers/risk-management/building-science/safe-rooms/funding.

Chapter 7

1 Gates, A.E., *Earth's Fury: Science of Natural Disasters* (Hoboken, NJ: Wiley & Sons-Blackwell Publishing, 2022), 403.

2 Stiros, S.C., "The 8.5+ Magnitude, AD365 Earthquake in Crete: Coastal Uplift, Topography Changes, Archaeological and Historical Signature," *Quaternary International* 216, no. 1–2 (2010): 54–63.

3 Shaw, B., N.N. Ambraseys, P.C. England, M.A. Floyd, G.J. Gorman, T.F.G. Higham, J.A. Jackson, J.-M. Nocquet, C.C. Pain, and M.D. Piggott, "Eastern Mediterranean

Tectonics and Tsunami Hazard Inferred from the AD 365 Earthquake," *Nature Geoscience* 1, no. 8 (2008): 268–76.

4 Ott, R.F., Karl W. Wegmann, Sean F. Gallen, Frank J. Pazzaglia, Mark T. Brandon, Kosuke Ueda, and Charalampos Fassoulas, "Reassessing Eastern Mediterranean Tectonics and Earthquake Hazard from the 365 CE Earthquake," *AGU Advances* 2, no. 2 (2021). doi:10.1029/2020AV000315.

5 Salama, A., M. Meghraoui, M. El Gabry, S. Maouche, M.H. Hussein, and I. Korrat, "Paleotsunami Deposits along the Coast of Egypt Correlate with Historical Earthquake Records of Eastern Mediterranean," *Natural Hazards and Earth System Sciences* 18 (2018): 2203–19.

6 Pararas-Carayannis, G., "The Earthquake and Tsunami of July 21, 365 AD in the Eastern Mediterranean Sea—Review of Impact on the Ancient World—Assessment of Recurrence and Future Impact," *Science of Tsunami Hazards* 30, no. 4 (2011): 253–92.

7 Causse, M., Emeline Maufroy, Lucile André, and Pierre-Yves Bard, "What Was the Level of Ground Motion across Europe during the Great AD 365 Crete Earthquake?" *Seismological Research Letters* 94, no. 5 (2023): 2397–410.

8 Ott, Wegmann, Gallen, Pazzaglia, Brandon, Ueda, and Fassoulas, "Reassessing Eastern Mediterranean Tectonics."

9 Pararas-Carayannis, "The Earthquake and Tsunami of July 21, 365 AD in the Eastern Mediterranean Sea."

10 Salama, Meghraoui, El Gabry, Maouche, Hussein, and Korrat, "Paleotsunami Deposits along the Coast of Egypt."

11 Werner, V., Kalliopi Baika, Peter Fischer, Hanna Hadler, Lea Obrocki, Timo Willershäuser, Anastasia Tzigounaki, et al., "The Sedimentary and Geomorphological Imprint of the AD 365 Tsunami on the Coasts of Southwestern Crete (Greece)—Examples from Sougia and Palaiochora," *Quaternary International* 473 (2018): 66–90.

12 Bahrouni, N., Mustapha Meghraoui, Hafize Başak Bayraktar, Stefano Lorito, Mohamed Fawzi Zagrarni, and Nabil Bel Mabrouk, "Evidence of Tsunami Deposits in East Tunisia Coastline Contemporaneous of the AD 365 Crete Earthquake: Field Data and Modelling," in *EGU General Assembly Conference*, 2021, 9104.

13 Polonia, A., Enrico Bonatti, Angelo Camerlenghi, Renata Giulia Lucchi, Giuliana Panieri, and Luca Gasperini, "Mediterranean Megaturbidite Triggered by the AD 365 Crete Earthquake and Tsunami," *Scientific Reports* 3, no. 1 (2013): 1285.

14 Gates, A.E. and D. Ritchie, *E-book: Encyclopedia of Earthquakes and Volcanoes* (New York: Facts On File, Inc., 2016), 368.

15 Pararas-Carayannis, "The Earthquake and Tsunami of July 21, 365 AD in the Eastern Mediterranean Sea."

16 Stiros, S.C., "Was Alexandria (Egypt) Destroyed in Ad 365? A Famous Historical Tsunami Revisited," *Seismological Society of America* 91, no. 5 (2020): 2662–73.

17 Stiros, "The 8.5+ Magnitude, AD365 Earthquake in Crete."

18 Pararas-Carayannis, "The Earthquake and Tsunami of July 21, 365 AD in the Eastern Mediterranean Sea."

19 Gates and Ritchie, *E-book*.

20 Satake, K., "Advances in Earthquake and Tsunami Sciences and Disaster Risk Reduction Since the 2004 Indian Ocean Tsunami," *Geoscience Letters* 1 (2014): 1–13.

21 Gates and Ritchie, *E-book*.

22 Ioualalen, M., Jack Asavanant, N. Kaewbanjak, S.T. Grilli, J.T. Kirby, and Phil Watts, "Modeling the 26 December 2004 Indian Ocean Tsunami: Case Study of Impact in Thailand," *Journal of Geophysical Research: Oceans* 112, no. C7 (2007). doi:10.1029/2006JC003850.

23 Science on a Sphere, "Tsunami Wave Run-Ups: Indian Ocean—2004," March 31, 2015 [cited 3/24/2025], https://sos.noaa.gov/catalog/datasets/tsunami-wave-run-ups-indian-ocean-2004/.

24 Ioualalen, Asavanant, Kaewbanjak, Grilli, Kirby, and Watts, "Modeling the 26 December 2004 Indian Ocean Tsunami."

25 "Girl, 10, Used Geography Lesson to Save Lives," *The Telegraph*, January 1, 2005, https://www.telegraph.co.uk/news/1480192/Girl-10-used-geography-lesson-to-save-lives.html.

26 Goff, J., Philip L.-F. Liu, Bretwood Higman, Robert Morton, Bruce E. Jaffe, Harindra Fernando, Patrick Lynett, Hermann Fritz, Costas Synolakis, and Starin Fernando, "Sri Lanka Field Survey after the December 2004 Indian Ocean Tsunami," *Earthquake Spectra* 22, no. S3 (2006): 155–72.

27 Science on a Sphere, "Tsunami Wave Run-Ups."

28 Carey, B., "Tsunami Waves Channeled Around the Globe in 2004 Disaster," *LiveScience*, 2005.

29 National Center for Environmental Information, *Sumatra, Indonesia Earthquake and Tsunami* (Washington, DC: NOAA, December 26, 2004).

30 Satake, K., "Advances in Earthquake and Tsunami Sciences and Disaster Risk Reduction Since the 2004 Indian Ocean Tsunami."

31 Billon, P.L., and Arno Waizenegger, "Peace in the Wake of Disaster? Secessionist Conflicts and the 2004 Indian Ocean Tsunami," *Transactions of the Institute of British Geographers* 32, no. 3 (2007): 411–27.

Chapter 8

1 Burt, J.E., and E. Aguado, *Understanding Weather & Climate, Global Edition*, Seventh Edition (New York: Pearson, 2015), 608.

2 Rippey, B., *U.S. Drought Monitor* (Lincoln, NB: National Drought Mitigation Center), https://droughtmonitor.unl.edu/.

3 National Center for Environmental Information, "U.S. Climate Extremes Index (CEI)," *NOAA*, https://www.ncei.noaa.gov/access/monitoring/cei/.

4 Gates, A.E., *Earth's Fury: Science of Natural Disasters* (Hoboken, NJ: Wiley & Sons-Blackwell Publishing, 2022), 403.

5 Williams, A.P., et al., "Large Contribution from Anthropogenic Warming to an Emerging North American Megadrought," *Science* 368, no. 6488 (2020): 314–18.

6 Sharer, R.J., and Loa P. Traxler, *The Ancient Maya*, Sixth Edition (Stanford, CA: Stanford University Press, 2006), 931.

7 Coe, M.D., *Breaking the Maya Code*, Third Edition (London: Thames & Hudson, 2012).

8 Kennett, D.J., M. Masson, C.P. Lope, et al., "Drought-Induced Civil Conflict Among the Ancient Maya," *Nature Communications* 13 (2022): 3911.

9 National Center for Environmental Information, *Drought and the Ancient Maya Civilization*, October 2021 [cited 2025 3/23/2025], https://www.ncei.noaa.gov/sites/default/files/2021-11/7%20Drought%20and%20the%20Ancient%20Maya%20Civilization%20-%20FINAL%20OCT%202021.pdf.

10 Golden, C., Andrew K. Scherer, Whittaker Schroder, Timothy Murtha, Shanti Morell-Hart, Juan Carlos Fernandez Diaz, Socorro del Pilar Jiménez Álvarez, Omar Alcover Firpi, Mark Agostini, Alexandra Bazarsky, et al., "Airborne Lidar Survey, Density-Based Clustering, and Ancient Maya Settlement in the Upper Usumacinta River Region of Mexico and Guatemala," *Remote Sensing* 13, no. 20 (2021): 4109.

11 Dunning, N., T. Beach, and S. Luzzadder-Beach, "Ancient Maya Agriculture," in *The Maya World* (Routledge, 2020), 501–18.

12 Sharer and Traxler, *The Ancient Maya*.

13 Stromberg, J., "Why Did the Mayan Civilization Collapse? A New Study Points to Deforestation and Climate Change," *Smithsonian Magazine*, 2012, https://www.smithsonianmag.com/science-nature/why-did-the-mayan-civilization-collapse-a-new-study-points-to-deforestation-and-climate-change-30863026/.

14 Golden, C., Andrew K. Scherer, Stephen Houston, Whittaker Schroder, Shanti Morell-Hart, Socorro del Pilar Jiménez Álvarez, George Van Kollias, et al., "Centering

the Classic Maya Kingdom of Sak Tz'i'," *Journal of Field Archaeology* 45, no. 2 (2020): 67–85.

15 Earth Observatory, "Mayan Deforestation and Drought," 2012 [cited 2025 3/23/2025], https://earthobservatory.nasa.gov/images/77060/mayan-deforestation-and-drought.

16 Gill, R.B., Paul A. Mayewski, Johan Nyberg, Gerald H. Haug, and Larry C. Peterson, "Drought and the Maya Collapse," *Ancient Mesoamerica* 18, no. 2 (2007): 283–302.

17 Golden, Scherer, Schroder, Murtha, Morell-Hart, Fernandez Diaz, del Pilar Jiménez Álvarez, Alcover Firpi, Agostini, Bazarsky, et al., "Airborne Lidar Survey, Density-Based Clustering, and Ancient Maya Settlement."

18 Kennett, D.J., M. Masson, C.P. Lope, et al., "Drought-Induced Civil Conflict Among the Ancient Maya," *Nature Communications* 13 (2022): 3911.

19 Fedick, S.L., and L.S. Santiago, "Large Variation in Availability of Maya Food Plant Sources during Ancient Droughts," *PNAS* 119, no. 1 (2021): e2115657118.

20 Stromberg, "Why Did the Mayan Civilization Collapse?"

21 Gates, *Earth's Fury*.

22 Xiao, L., Xiuqi Fang, Jingyun Zheng, and Wanyi Zhao, "Famine, Migration and War: Comparison of Climate Change Impacts and Social Responses in North China between the Late Ming and Late Qing Dynasties," *The Holocene* 25, no. 6 (2015): 900–10.

23 Li, K., *Climate Change, Epidemics, Fiscal Breakdown, and the Collapse of the Great Ming* (University of Michigan), https://osf.io/rwe32/download.

24 Ditmanson, P.B., "Ming Dynasty," in *Berkshire Encyclopedia of China*, ed. L. Cheng (Berkshire Publishing Group, 2009), 1483–89.

25 Zhao, J., et al., "Role of the Summer Monsoon Variability in the Collapse of the Ming Dynasty: Evidences from Speleothem Records," *Geophysical Research Letters* 48, no. 11 (2021): e2021GL093071.

26 Fan, K.-W., "The Little Ice Age and the Fall of the Ming Dynasty: A Review," *Climate* 11, no. 3 (2023): 71.

27 Brook, T., "Nine Sloughs: Profiling the Climate History of the Yuan and Ming Dynasties, 1260–1644," *Journal of Chinese History* 1, no. 1 (2017): 27–58.

28 Xiao, Fang, Zheng, and Zhao, "Famine, Migration and War."

29 Zhao, et al., "Role of the Summer Monsoon Variability in the Collapse of the Ming Dynasty."

30 Tong, J.W., *Disorder under Heaven: Collective Violence in the Ming Dynasty* (Stanford, CA: Stanford University Press, 1992).

31 Brook, T., "Telling Famine Stories: The Wanli Emperor and the 'Henan Famine' of 1594," *Études chinoises. 漢學研究* 34, no. 2 (2015): 163–202.

32 Xiao, Fang, Zheng, and Zhao, "Famine, Migration and War."

33 Tong, *Disorder under Heaven*.

34 Storozum, M., Peng Lu, Sanying Wang, Panpan Chen, Ruixia Yang, Qifeng Ge, Jinping Cao, et al., "Geoarchaeological Evidence of the AD 1642 Yellow River Flood That Destroyed Kaifeng, a Former Capital of Dynastic China," *Scientific Reports* 10, no. 1 (2020): 3765.

35 Chen, K., et al., "One Drought and One Volcanic Eruption Influenced the History of China: The Late Ming Dynasty Mega-Drought," *Geophysical Research Letters* 47, no. 16 (2020): e2020GL088124.

36 Li, Q.-H. et al., "Overview of the Plague in the Late Ming Dynasty and Its Prevention and Control Measures," *Traditional Medicine Research* 5, no. 3 (2020): 136–44.

37 Li, et al., "Overview of the Plague in the Late Ming Dynasty."

38 Fan, "The Little Ice Age and the Fall of the Ming Dynasty."

39 Xiao, Fang, Zheng, and Zhao, "Famine, Migration and War."

40 Fan, "The Little Ice Age and the Fall of the Ming Dynasty."

41 Gates, *Earth's Fury*.

42 Library of Congress, "The Dust Bowl," Presentation: U.S. History Primary Source Timeline [cited 3/23/2025], https://www.loc.gov/classroom-materials/united-states-history-primary-source-timeline/great-depression-and-world-war-ii-1929-1945/dust-bowl/.

43 Cook, B.I., R.L. Miller, and R. Seager, "Amplification of the North American 'Dust Bowl' Drought through Human-Induced Land Degradation," *Proceedings of the National Academy of Sciences* 106, no. 13 (2009): 4997–5001.

44 Cook, Miller, and Seager, "Amplification of the North American 'Dust Bowl' Drought."

45 Baumhardt, R.L., "Dust Bowl Era," *Encyclopedia of Water Science*, 2003, 491.

46 Cook, Miller, and Seager, "Amplification of the North American 'Dust Bowl' Drought."

47 Burnette, D.J., and D.W. Stahle, "Historical Perspective on the Dust Bowl Drought in the Central United States," *Climatic Change* 116 (2013): 479–94.

48 Cook, Miller, and Seager, "Amplification of the North American 'Dust Bowl' Drought."

49 Baveye, P.C., et al., "From Dust Bowl to Dust Bowl: Soils Are Still Very Much a Frontier of Science," *Soil Science Society of America Journal* 75, no. 6 (2011): 2037–48.

50 "The Black Sunday Dust Storm of April 14, 1935," National Weather Service [cited 3/24/2025], https://www.weather.gov/oun/events-19350414.

51 History.com Editors, "Dust Bowl," October 27, 2009 [cited 3/23/2025], https://www.history.com/topics/great-depression/dust-bowl.

52 Worster, D., *Dust Bowl: The Southern Plains in the 1930s 25th Anniversary Edition* (New York: Oxford University Press, 2004), 304.

53 Gates, *Earth's Fury.*

Chapter 9

1 Burt, J.E., and E. Aguado, *Understanding Weather & Climate, Global Edition*, Seventh Edition (New York: Pearson, 2015), 608.

2 Gates, A.E., *Earth's Fury: Science of Natural Disasters* (Hoboken, NJ: Wiley & Sons-Blackwell Publishing, 2022), 403.

3 Kessler, B., "Doggerland: The Europe That Was," *Education* [cited 3/23/2025], https://education.nationalgeographic.org/resource/doggerland/.

4 Vučković, A., "St. Lucia's Flood: The Disaster That Changed the Shape of Europe," *Ancient Origins*, March 25, 2023 [cited 3/24/2025], https://www.ancient-origins.net/history-important-events/st-lucias-flood-0018127.

5 Eaton, G., *A History of Civilization in 50 Disasters (History in 50)* (Tilbury House Publishers and Cadent Publishing, 2015).

6 Ossing, F., Jörg FW Negendank, and Rolf Emmermann, "The Origin of Landscape," in *Die Kleine Eiszeit. Holländische Landschaftsmalerei im 17* (Gemäldegalerie der Staat Museen zu Berlin, Editor, 2001), 26–40.

7 Knottnerus, O., "Reclamations and Submerged Lands in the Ems River Estuary (900–1500)," *Landscapes or Seascapes* (2013): 241–66. doi:10.1484/M.CORN.1.101555.

8 Ossing, Negendank, and Emmermann, "The Origin of Landscape."

9 Eaton, *A History of Civilization in 50 Disasters.*

10 Vučković, "St. Lucia's Flood."

11 Page, W., "Houses of Austin Canons: The Priory of Hickling," in *A History of the County of Norfolk* (London: British History Online, 1906), 383–6.

12 Vučković, "St. Lucia's Flood."

13 Haslett, S.K., and E.A. Bryant, "Reconnaissance of Historic (Oost-AD 1000) High-Energy Deposits along the Atlantic Coasts of Southwest Britain, Ireland and Brittany, France," *Marine Geology* 242, no. 1–3 (2007): 207–20.

14 Hambling, D., "Weatherwatch: St Lucia's Storm, the Flood that Changed the Face of Europe," *Guardian*, December 12, 2018 [cited 3/24/2025], https://www.theguardian.com/news/2018/dec/12/st-lucias-storm-the-flood-that-changed-the-face-of-netherlands-europe.

15 Vučković, "St. Lucia's Flood."

16 Spencer, R., "Contested Terrain: The Mississippi Flood of 1927 and the Struggle to Control Black Labor," *The Journal of Negro History* 79, no. 2 (1994): 170–81.

17 Hornbeck, R., and S. Naidu, "When the Levee Breaks: Black Migration and Economic Development in the American South," *American Economic Review* 104, no. 3 (2014): 963–90.

18 Rivera, J. D., and D. S. Miller, "Continually Neglected: Situating Natural Disasters in the African American Experience," *Journal of Black Studies* 37, no. 4 (2007): 502–22.

19 Spencer, "Contested Terrain."

20 Barry, J.M., *Rising Tide: The Great Mississippi Flood of 1927 and How It Changed America* (New York: Simon & Schuster, 2007).

21 Ambrose, S., "Man vs. Nature: The Great Mississippi Flood of 1927," *National Geographic*, May 1, 2001, https://www.nationalgeographic.com/culture/article/mississippi-river-flood-culture.

22 Barry, *Rising Tide.*

23 Parrish, S.S., *The Flood Year 1927: A Cultural History* (Princeton, NJ: Princeton University Press, 2007).

24 Ambrose, "Man vs. Nature."

25 Spencer, "Contested Terrain."

26 Hornbeck and Naidu, "When the Levee Breaks."

27 Lemann, N., *The Promised Land: The Great Black Migration and How It Changed America* (New York: Vintage, 1992).

28 Parrish, *The Flood Year 1927.*

29 Barry, *Rising Tide.*

30 Gates, *Earth's Fury.*

31 Greer, C.E., and Igor Vladimirovich Popoov, "Yellow River," in *Encyclopedia Britannica* (Britannica, 1999).

32 Muscolino, M.S., *The Ecology of War in China: Henan Province, the Yellow River, and Beyond, 1938–1950* (Cambridge: Cambridge University Press, 2014).

33 Lary, D., "Drowned Earth: The Strategic Breaching of the Yellow River Dyke, 1938," *War in History* 8, no. 2 (2001): 191–207.

34 Edgerton-Tarpley, K.J., "From 'Nourish the People' to 'Sacrifice for the Nation': Changing Responses to Disaster in Late Imperial and Modern China," *The Journal of Asian Studies* 73, no. 2 (2014): 447–69.

35 Muscolino, M.S., "Violence Against People and the Land: The Environment and Refugee Migration from China's Henan Province, 1938–1945," *Environment and History* 17, no. 2 (2011): 291–311.

36 Edgerton-Tarpley, "From 'Nourish the People' to 'Sacrifice for the Nation.'"

Chapter 10

1 Ritchie, H., Lucas Rodés-Guirao, Edouard Mathieu, Marcel Gerber, Esteban Ortiz-Ospina, Joe Hasell and Max Roser, "Population Growth," *OurWorldinData.org*, 2023.

2 IRIS, "Fact Sheet: How Often Do Earthquakes Occur," Education and Outreach Series [cited 3/25/2025], https://www.iris.edu/hq/inclass/fact-sheet/how_often_do_earthquakes_occur.

3 Earthquake Hazards Program, "M 9.5—1960 Great Chilean Earthquake (Valdivia Earthquake)," USGS [cited 3/24/2025], https://earthquake.usgs.gov/earthquakes/eventpage/official19600522191120_30/impact.

4 Satake, K., and B.F. Atwater, "Long-Term Perspectives on Giant Earthquakes and Tsunamis at Subduction Zones," *Annual Review of Earth and Planetary Sciences* 35, no. 1 (2007): 349–74.

5 Gates, A.E., and D. Ritchie, *E-book: Encyclopedia of Earthquakes and Volcanoes* (New York: Facts on File, Inc., 2016), 368.

6 Gates, A.E., *Earth's Fury: Science of Natural Disasters* (Hoboken, NJ: Wiley & Sons-Blackwell Publishing, 2022), 403.

7 Minor, R., and W.C. Grant, "Earthquake-Induced Subsidence and Burial of Late Holocene Archaeological Sites, Northern Oregon Coast," *American Antiquity* 61, no. 4 (1996): 772–81.

8 Clague, J.J., "Evidence for Large Earthquakes at the Cascadia Subduction Zone," *Reviews of Geophysics* 35, no. 4 (1997): 439–60.

9 Gates, *Earth's Fury.*

10 Hough, S.E., "Scientific Overview and Historical Context of the 1811–1812 New Madrid Earthquake Sequence," *Annals of Geophysics* 47, no. 2–3 (2004): 523–37.

11 Johnston, A.C., and E.S. Schweig, "The Enigma of the New Madrid Earthquakes of 1811–1812," *Annual Review of Earth and Planetary Sciences* 24, no. 1 (1996): 339–84.

12 Hough, "Scientific Overview and Historical Context of the 1811–1812 New Madrid Earthquake Sequence."

13 Bagnall, N.H., *On Shaky Ground: The New Madrid Earthquakes of 1811–1812*. Vol. 1 (Columbia, MO: University of Missouri Press, 2013).

14 Gates and Ritchie, *E-book.*

15 Gates, *Earth's Fury.*

16 Fritz, H.M., Fahad Mohammed, and Jeseon Yoo, "Lituya Bay Landslide Impact Generated Mega-Tsunami 50th Anniversary," *Pure and Applied Geophysics* 166 (2009): 153–75.

17 Gates, *Earth's Fury.*

18 Ward, S.N., and S. Day, "Cumbre Vieja Volcano—Potential Collapse and Tsunami at La Palma, Canary Islands," *Geophysical Research Letters* 28, no. 17 (2001): 3397–400.

19 Hawaiian Volcano Observatory, "Volcano Watch—The Canary Islands 'Mega-Tsunami' Hypothesis, and Why It Doesn't Carry Water," October 21, 2021 [cited 3/24/2025], https://www.usgs.gov/observatories/hvo/news/volcano-watch-canary-islands-mega-tsunami-hypothesis-and-why-it-doesnt-carry.

20 Gates, *Earth's Fury.*

21 Buchholz, K., "2010s See Record Number of Storms in the Philippines," *Tropical Storms*, 2020, Statista.

22 Mas, E., et al., "Field Survey Report and Satellite Image Interpretation of the 2013 Super Typhoon Haiyan in the Philippines," *Natural Hazards and Earth System Sciences* 15, no. 4 (2015): 805–16.

23 Singer, M. "2013 State of the Climate: Record-Breaking Super Typhoon Haiyan," *Climate.gov*, 2014.

24 Bhatia, K., et al., "A Potential Explanation for the Global Increase in Tropical Cyclone Rapid Intensification," *Nature Communications* 13, no. 1 (2022): 6626.

25 Radfar, S., H. Moftakhari, and H. Moradkhani, "Rapid Intensification of Tropical Cyclones in the Gulf of Mexico Is More Likely During Marine Heatwaves. *Communications Earth & Environment* 5, no. 1 (2024): 421.

26 Pasch, R.J., Andrew B. Penny, and Robbie Berg, "Hurricane Maria," in *Tropical Cyclone Report* (National Hurricane Center, 2023), 48.

27 Greenemeier, L., "Repair or Renovate? Puerto Rico Faces Stark Power Grid Options," *Scientific American*, November 13, 2017, https://www.scientificamerican.com/article/repair-or-renovate-puerto-rico-faces-stark-power-grid-options/.

28 National Weather Service, "Hurricane Helene Makes Landfall in the Florida Big Bend September 26–27, 2024," [cited 3/23/2025], https://www.weather.gov/tae/helene2024.

29 National Severe Storms Laboratory, "2011 Spring Tornado Outbreaks," [cited 3/23/2025], https://www.nssl.noaa.gov/about/history/2011/.

30 Gates, *Earth's Fury*.

31 Simmons, K., and Daniel Sutter, *Deadly Season: Analysis of the 2011 Tornado Outbreaks* (Boston, MA: American Meteorological Society, 2012).

32 National Severe Storms Laboratory, "2011 Spring Tornado Outbreaks."

33 Simmons and Sutter, *Deadly Season*.

34 Gates, *Earth's Fury*.

35 Smith, R.B., and L.W. Braile, "The Yellowstone Hotspot," *Journal of Volcanology and Geothermal Research* 61, no. 3–4 (1994): 121–87.

36 Breining, G., *Super Volcano: The Ticking Time Bomb Beneath Yellowstone National Park* (Voyageur Press, 2007).

37 Uri, J., "115 Years Ago: The Tunguska Asteroid Impact Event," NASA, June 30, 2023 [cited 2025 3/23/2025], https://www.nasa.gov/history/115-years-ago-the-tunguska-asteroid-impact-event/.

38 Morgan, J.V., et al., "The Chicxulub Impact and Its Environmental Consequences," *Nature Reviews Earth & Environment* 3, no. 5 (2022): 338–54.

39 Range, M.M., et al., "The Chicxulub Impact Produced a Powerful Global Tsunami," *Agu Advances* 3, no. 5 (2022): e2021AV000627.

40 Morgan, et al., "The Chicxulub Impact and Its Environmental Consequences."

41 Schulte, P., et al., "The Chicxulub Asteroid Impact and Mass Extinction at the Cretaceous-Paleogene Boundary," *Science* 327, no. 5970 (2010): 1214–18.

INDEX

About the Author

Alexander Gates is a distinguished service professor of earth and environmental sciences at Rutgers University in Newark, New Jersey. Dr. Gates holds a PhD in Geology from Virginia Tech specializing in faults. He has published twelve books and eighty-four professional papers and has edited eleven professional volumes. His work in geology, education, and outreach has been recognized with twenty-nine professional awards. He has also appeared on numerous television news, radio, newspaper, and magazine stories, as well as several documentaries on earthquakes and plate tectonics.